Water Quality and Systems: A Guide for Facility Managers

2nd Edition

Water Quality and Systems: A Guide for Facility Managers

2nd Edition

Robert N. Reid, PE

THE FAIRMONT PRESS, INC.
Lilburn, Georgia

MARCEL DEKKER, INC.
New York and Basel

Library of Congress Cataloging-in-Publication Data

Reid, Robert N., 1949-
    Water quality and systems : a guide for facility managers/Robert N. Reid.-
-2nd ed.
        p. cm.
    Includes bibliographical references and index.
    ISBN 0-88173-332-6 (print) -- ISBN 0-88173-463-2 (electronic)
    1. Plumbing. 2. Water quality management. 3. Water supply. I. Title

    TH6126.R32 2003
    696'.1--dc21

                                                        2003054958

Published by The Fairmont Press, Inc.
700 Indian Trail, Lilburn, GA 30047
tel: 770-925-9388; fax: 770-381-9865
http://www.fairmontpress.com

Distributed by Marcel Dekker, Inc.
270 Madison Avenue, New York, NY 10016
tel: 212-696-9000; fax: 212-685-4540
http://www.dekker.com

Printed in the United States of America

10   9   8   7   6   5   4   3   2   1

0-88173-332-6 (The Fairmont Press, Inc.)
0-8247-4010-6 (Marcel Dekker, Inc.)

While every effort is made to provide dependable information, the
publisher, authors, and editors cannot be held responsible for any
errors or omissions.

Dedication

This book is dedicated to Katherine Jane Schow Reid.

Contents

Preface

*W*ater Quality & Systems: A Guide for Facility Managers is written for facility managers and other building professionals—such as maintenance executives, consultants and technical professionals—who have been given the responsibility for water supply and wastewater drainage systems. This book is designed to help the facility manager perform the core job of integrating people into their physical environment.

Managers want to know how to improve their water system quality and how to save money on water and wastewater costs. This book provides simple distinct steps that enable managers to keep occupants satisfied and productive. At the same time, facility managers will learn how to control and maintain operating and repair costs and how to implement the requirements of new regulations into the systems.

For management professionals who are confused by technical lingo, this book cuts through the jargon and shows facility managers how to reduce costs and make water systems safe and efficient in a straightforward, non-technical reference. It has been specifically formatted to provide the manager with the most important, most comprehensive amount of information in the least amount of time, thereby optimizing the reader's investment in the information, not in the writing.

The material covered in this book will also be useful for technical professionals who want to be able to communicate more effectively on a non-technical level with their clients and customers. It will provide a bridge of communication between facility managers and technical staffs and/or consultants optimizing time and staff investment.

The text is divided into problem segments with immediate response. It is not a textbook designed to be covered over the standard 13-16 week course. Instead, it is designed to provide, at quick reading, the assessment of the problem, drive to the root of it, and quickly determine for the manager the actions to be taken to solve them in the near term and long term.

Water supply and wastewater are the two broad types discussed in this book. The basic important elements in each category are presented. The book addresses significant problems faced by managers of water systems. It also explains analysis of water supplies, treatment costs and methods of saving money in treatment, remodeling, construction and operations.

Managers of water systems will profit from the books current information about new and dynamic water regulations and will be able to pass this information along to their employees in order to operate the system effectively and safely.

At the completion of the book, the facility manager will have learned how to provide occupants with high quality water supply and wastewater systems and minimum costs without compromising safety. In addition, managers will be able to communicate effectively with semi-technical people about his or her water system. New and more effective methods of treatment, installation and operation will be learned.

The reader of this book will know how to keep water safe for its consumers-how to make drainage and sewer systems safely carry away waste. The reader will also understand how to test the systems, who to contact in order to check and verify the tests, how to determine the cost of improvements and how to analyze the costs to determine if they are effective. Lastly, managers will obtain knowledge of recent laws and regulations concerning water and wastewater systems.

Finally, the book can be used as a guide to increase awareness among future managers, the facility management staff and/or students studying water systems.

Introduction

PROBLEMS AND SOLUTIONS

The phone rings. The administrative assistant answers. Somewhere, suddenly, there is a leak, the water is off, or "the water tastes funny." Suddenly, the facility manager has a problem and if it is not solved quickly, top management is going to be calling and asking some hard questions.

Water Quality & Systems: A Guide for Facility Managers provides the most up-to-date, comprehensive information for today's facility manager and other building professionals such as maintenance personnel and consultants. It provides a series of management steps to be taken to successfully control and manage water supplies and wastewater systems.

NEW TECHNOLOGIES

Water managers continue to face new challenges in light of changes in technology, in monitoring, in materials and in regulatory requirements. New technologies are being applied to supply and purify water sources and to remove unwanted elements. The information in this text will increase the facility manager's knowledge and appreciation of the subject but it will also provide insight into safe, cost-effective management as well.

PLANNED WATER MANAGEMENT

Successful water managers know three things about their water. They know where their water comes from; they know what chemicals are in it; they know how much it costs to get it and pay for treatment after they have used it. This book shows managers how water is successfully used and managed and how rules and regulations are changing in the field of water quality and treatment.

WATER IN, WATER OUT

Water management is divided into two fields. Water coming into a plant is pure and some portion of it is used for drinking and bathing. People assume, without question, that this water is safe. It is a basic philosophy. In order for this assumption to remain valid, managers and staff work harder and harder to maintain quality because purity of water supplies are slowly declining.

Once the water has been used, it becomes wastewater and is sent, through pipes and networks, to a plant where it is treated (i.e., cleaned) and discharged. It can go into another body of water, onto the ground, or into ponds until it evaporates and leaves the remaining impurities behind.

Successful use of water depends upon the knowledge of the people managing it and recognizing the importance it plays upon the clients or customers for whom it is supplied.

In reality, the water from one facility becomes the water in to the next. Once the concept of this cycle is grasped, the importance of good water management becomes clear.

In the past, water came from wells or streams and was surprisingly crystal clear and high quality. In processing, pollution of many forms was added and reduced the cleanliness. However, as there were no downstream users, or since the downstream users had their own quality supply, it did not matter how much the water became contaminated.

Today, water is treated and tested before being put into the network. Sophisticated techniques are used to determine what is in water before and after usage to confirm treatment remains effective. Several government agencies have increased authority in order to protect this precious resource. Their weight and mandate can certainly be oppressive if the perception is attained that the facility user is not a good steward. Readers will be able to avoid making potential costly mistakes with their water.

SUPPLY

Where should we get our water from? How much should it cost us? What systems and energies are used to get it to the facility? What uses does water have? Use is broken down for ready access by the reader. From there, this book provides insight into how it is moved, stored, treated, and some of the logic behind the way water is handled.

CONSERVATION

The book provides the facility manager with tips for water conservation and even tells hotel and motel facility managers and school managers how to get a free computer modeling program that allows him to perform trial and error alternative analysis on his own water system.

WASTEWATER

Besides supply water, this book also looks at the wastewater side of the equation, into the kinds of pipes and components used to haul water away. The key difference for the different type of designs for the wastes will be explained. Materials and components will be reviewed and clarified. Types of piping, pumping and the various methods of drainage and venting are addressed. All with subheads to provide quick access and reduce search and scan time.

In addition, the more sophisticated elements of water systems such as softeners and de-ionizers are examined and discussed. Forms and tables are provided to help the manager determine quickly and effectively the value of water softening.

Many materials, fittings, and theories common to both systems are addressed including sketches and diagrams. Case histories and examples are included.

REGULATIONS

New regulations that make the facility manager's job more challenging are discussed along with some helpful hints in dealing with regulators. The revised Safe Drinking Water Act is discussed along with the rules for public notification of impurities in supplies.

A discussion of the interface between the two types of systems will be presented. This will be dedicated to effective design of restroom and bathing facilities and cases of successful layouts will be documented.

WATER PURITY

A complete chapter on water purity is provided that discusses, among other things, various forms of impurities and the treatment methods for removing them. A table of health hazards and regulatory standards is provided to indicate to the facility manager what the requirements are. The chapter tells a facility manager how to hire a laboratory, how to have water quality tests conducted and how to interpret the lab's results.

CONSTRUCTION

Methods of construction, along with the tradeoffs, will be shown. Sources of pricing for labor and materials and computation of quantities will be provided. Forms and formats are included to make the manager's duties simpler and more effective.

Finally, the text will go into the testing that accompanies any major system modification or retrofit. The testing will be based upon accepted industry standards.

Appendices at the end of the text will furnish guides to industry associations and manufacturers.

Throughout *Water Quality & Systems* are many cost-saving, money-saving ideas or tips that will make it well worth reading. The first step—water supply, or getting water where you need it when you need it—begins in Chapter 1.

Chapter 1

Water Management

Planned water management can be a challenging task for facility managers. It requires managing the water supplies and wastewaters into and out of his facility, and recognizing the significant costs and risks. An understanding of the basic elements of successful water management is an essential building block to ensuring that the proper quantity and quality of water is available every day to occupants and visitors.

PLANNED WATER MANAGEMENT

Just last week, the local press reported sewage had been discovered in the drinking water system of the little town across the valley. The city officials were sterilizing the system but citizens had been advised not to drink the water and to sterilize it by boiling it before using it.

No water manager or professional wants this to happen. It reflects upon the care and trust of people who rely on water for life. Successful water managers make sure their water is clean. Codes, standards and laws are written to make sure it stays clean. Water should be available in ample quantity and at the right pressure for everyone, because everyone needs it.

In addition, waste water, because of its potential hazards to health, should be carried safely to a point where it can be successfully treated and its pollutants safely and efficiently removed.

Within these broad goals, a successful water manager attempts to keep the costs of maintaining water quality, distribution and waste under control. Water and wastewater costs are heavily subsidized with tax dollars but as tax dollars dwindle, the burden of paying for water and water system management falls upon the facility.

Water In, Water Out

Unless the facility has taken upon itself the role of drinking water treatment, the water supply is assumed to be fit for drinking when it enters the facility. Most plants have a ready source of treated water, but after the water crosses the boundary into the facility the responsibility for maintaining the water's quality is the facility manager's.

A few communities in America use untreated supplies from deep wells. In many cases, the water is filtered and treated to remove disease-carrying microorganisms.

Once the water has been used, wastewater flows into a centralized collection system where it is conveyed to a treatment plant for removal of the pollutants. The treated water is then discharged into a lake, stream, canal or other body of water.

Water, therefore, is not actually consumed—it is simply changed from drinking water to sewer water as it passes through the facility.

WATER SUPPLIES

Most people do not worry about their water supply. Worry is the job of city officials. But local governments know where their water comes from and have projections about the amounts they have or will have available from year to year. Today, water use is changing and efforts are underway to reduce consumption wherever and whenever possible.

Water Costs

Water managers find their water is inexpensive. In fact, it is so inexpensive that most managers do not really worry about their budget when it comes to water costs. It is usually the costs of water treatment that surprise them. The fact is, a lot of tax dollars have already gone into the water supply. Most of the utility costs—the treatment plant, the water mains, the staff to do the treatment, chemical analysis, etc.—have already been taken care of by the tax-funded utility. The facility manager can rely upon good quality water but he pays for it every month, some paid for in the property tax, the balance paid for in the form of a water bill.

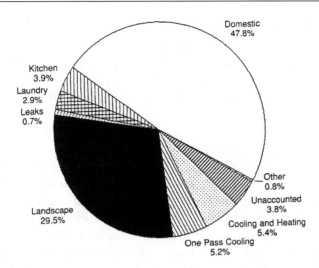

Figure 1-1. Typical water use. Courtesy: The Denver Board of Water Commissioners, Non-Residential Water Audit, Summary Report 1991.

QUANTITY

Individual Uses

Facility managers make use of water the same way residences do-in fact, most facility use is fairly parallel to home use. Water is used for washing, bathing, sanitary waste, cooking and drinking. In addition, facility managers use water for make-up of evaporation in pools, for cooling, in air conditioning, for lawns, gardening or grounds keeping. Managers also use water in industrial processes unique to their facility (see Figure 1-1).

For individuals, water requirements are going to be between 70 and 120 gallons per person per day. About 70 gallons per person per day is used for washing/bathing and for the toilet.

Office, Light Industrial, and Manufacturing Uses

In an office environment, with cooking and bathing left at home, water for personal use will be lower than the figures above.

Industrial and manufacturing facilities use water for cooling, washing, carrying away waste, and mixing.

Lawns and Agricultural Use

The other parameters, for lawns and grounds-keeping, vary with seasons, and with size of the complex. Lawns of grasses will require approximately eight/tenths of a gallon per day per square yard (3,500 gallons per acre) and are also dependent upon recent weather conditions.

For agricultural use, water usage depends upon season and the crop. Many sophisticated techniques are used to estimate plant demands for agriculture. For lawns, the most effective application is to use water sprinklers. For *agriculture*, sprinklers are also very effective but the sprinkler systems are quite expensive, and for this reason crops are often flood-irrigated. Flood irrigation is not a relatively effective use of water since much of the water is consumed by evaporating from the soil before it reaches the plants. In desert environments, a new method of water conservation, called drip irrigation, is used to apply water directly to the plant. Drip irrigation is approximately the same cost as using sprinklers.

Industrial Uses

Determining the needs for industrial processes is more complex, although engineers have become adept at estimating needs for several types of use. Where water is needed for cooling, engineers can figure out how much water is needed from knowing how much heat is generated or fuel is being burned. The heat is a result of the process calculations (for a short discussion of heat, see Chapter 12, hot water systems). Water use for mixing is computed by engineers in a similar way.

One major factor in calculating the sizes of the pipes for any facility is the need to provide water for fire-fighting. As a result, pipes are sized greater than needed to put out a fire.

DELIVERY

In addition to knowing the amounts to be delivered, the problem of delivering at the correct pressure is also encountered. While a few systems have natural pressure, most water supplies are pumped. Facilities can receive supply at a central metering station or at a series of stations located around the perimeter.

Water must have enough pressure to reach the point of delivery and must flow from the point of delivery in sufficient quantity to meet the demand.

For tall buildings or facilities located on a hill, the water is pressurized through the use of pumps and occasional storage in the upper stories of the buildings or water towers. Pumping to storage tanks allows use of smaller pumps. However, the facility must pay additional for the costs of the tanks for storage. (For a discussion of the tradeoffs between constant pumping and storage, see Chapter 8 where we discuss pipe hydraulics.)

POLLUTION / WASHING / WASTE

Management of wastewater from washing and carrying away wastes is different from the management of fresh water supplies. As with fresh water, pipes are used to carry the flows, and the wastewaters are derived from the incoming supplies. But wastewater management deals with essentially contaminated or polluted water, and hence the procedures and management principles are slightly different.

One overriding principal rule applied by environmental regulators is: "Dilution is not the solution to pollution."

A simple statement perhaps, but this statement, which reflects some strict environmental rules relative to hazardous waste and to a lesser extent to other types of wastes, means that washing away a spill or cleanup from processing operations is not generally allowed. The facility manager is required to clean up the water before letting it go back into the environment.

A case study is included at the end of this chapter. (For a discussion of the most significant environmental regulations faced by today's facility managers, see Chapter 3. For a complete discussion of wastewater management, see Chapter 13.)

UTILITY BILLINGS

Depending upon who the facility manager pays for water, the utility usually reads the meter and bills the user monthly. Past evaluation of utility billings is an excellent method of determining the quantity of water used throughout the year. Some utilities have started quarterly billings to save staff costs of reading meters and preparing the paperwork that goes along with the bills.

By evaluating the facility utility bills, the facility manager can determine trends in water use over the year. Facility mangers familiar with use of Personal Computers can use a database or spreadsheet program to plot uses over time from the billings (see Table 1-1 and the accompanying Figure 1-2 to see how past utility billings can aid in an assessment of quantity uses and demands).

Where in the United States or Canada the facility is located will determine the amount of information the facility manager can obtain from his monthly or quarterly billings. In some areas, the sewer billing is a function of the water supply which is done to eliminate costly metering of the sewage flows. In other areas, sewer billing is a flat fee depending upon the size of the complex or number of people working there. Some large facilities have their own water supply and sewage treatment systems.

These large self-contained facilities will be discussed in later chapters, but facility managers with their own systems should have metering and utility data in a similar manner to any utility. By metering and keeping records, the facility manager can report the value of the services being provided. This type of information is necessary because of the tendency to consider contracting out support services by many organizations.

Some terms used by engineers and utilities for measuring water use for large users include, cubic meters, cubic feet, gallons, and acre feet.

METERING

Metering the supply is usually the responsibility of the utility. Meters are typically located at the property boundary, and large users' meters are installed either underground in a vault or on the main water pipeline where it enters the building. Flow measurement meters work on the basis of one or two theoretical principles. These theories have been checked time after time and have been essentially proven by the industry (see Chapter 10 for further

Table 1-1. Table and plot of water use from utility bills (see Figure 1-1).

Month	Use In Gallons
January	13,000
February	12,000
March	18,000
April	22,000
May	25,000
June	28,000
July	33,000
August	30,000
September	26,000
October	21,000
November	17,000
December	14,000

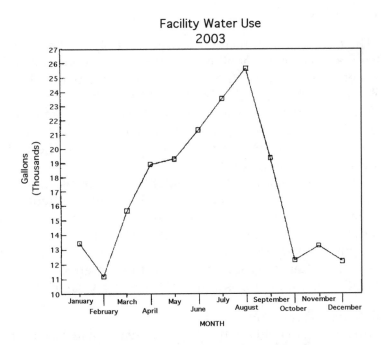

Figure 1-2. Facility water use 2003 (see Table 1-1).

discussion on meters).

In addition to metering the volume, the utility will sometimes provide a temporary meter that continuously monitors the
flow throughout the day for several days. This type of service is
potentially free since both the utility and the facility manager
benefit from the results of the information. The result of this type
service allows the facility manager to plot his or her peak flows
and enables him to locate major users.

In a large facility, multiple users of large amounts of water
can sometimes lead to unexpected problems. One facility found
that when the laundry was operating and the lawn sprinklers
were fully on, water pressure was so low in the high rise building
on the campus that the water coolers did not have enough pressure to allow students to get a drink. Spot pressurization, using
pumps, was necessary to keep the pressure up in one area while
use was curtailed in other areas. By measuring flows and pressures, the facility was able to isolate this root problem.

WATER TREATMENT

The most common method of water supply treatment in the
United States is the use of alum to settle out the solids and clarify
the water. A utility filters the water, adds the alum and allows it
to settle out, then injects chlorine to make sure the water is free of
bacteria. (New trends in water treatment using ozone and sophisticated disinfectants are explained thoroughly in Chapter 11.)

Once the water enters the property, however, the facility
manager is responsible for it.

A few facilities have their own supply. In these cases, the facility takes responsibility for performing its own treatment. Most
public supplies come from rivers or lakes, while some comes from
springs. Smaller communities get their supplies from wells where
less treatment is required. In general, well water is often hard and
needs to be softened. (Wells for water supplies are discussed in
Chapter 2. Water softening is a subject covered in Chapter 4).

Wastewater Treatment

Waste water is often required to be treated and facility managers who have added lots of chemicals in washing or used the

water for cooling or blanching are required by regulations to treat the water before it leaves the property. Such methods could include settlement as in a pond or tank and/or retention while micro-organisms remove organic material. (Other methods including filtering, aeration, percolation and digestion are discussed in Chapter 13.)

COMMON COMPLAINTS

Many first-time water managers find that water costs are such a low end of their budget they do not have to worry about cost when they take over their role as a facility manager. Energy (fuel) costs and electric power are first and second respectively in dollar value. But when there are water or wastewater system problems—the supply is too low, the water is off or seems cloudy, or the wastewater is contaminated with a pesticide—then the devil is due. Nobody cares how much it costs and nobody cares whether the facility manager saved any money in the account when the quality of the water has been questioned.

Common complaints include, "Gee the pressure is low," "Gee, the water tastes salty," "Gee, the water is cloudy... rusty... warm..." This and a host of other problems must be solved by the facility manger (see Table 1-3 for a list of common complaints and probable sources). One of the best things a manager can do is find out where the problem is and get a trained staff person there as quickly as possible to check the water and answer the complaint. In addition, water managers have come under fire recently for contaminated/or tainted supplies that, while they do not affect a healthy user, may potentially affect occupants who are immune-deficient in some way. This could include a person who has been diagnosed with HIV syndrome or someone with cancer, pneumonia or AIDS. These people, who would not normally be affected by minor amounts of impurities in the water supply, are likely to sue the facility that gave them impure water that exacerbates their illness.

Unspecified complaints should be verified with lab tests. Sometimes the source of the problem will be immediately apparent, but sometimes it disappears mysteriously. Too many complaints of water problems that are not isolated and solved by the facility manager will lead to problems for that manager in short order.

The only way a facility can adequately determine whether or not there is a water problem is by taking a sample of the water and having it analyzed in a laboratory (see Chapter 4).

LABORATORY TESTS

Since most facilities do not have a laboratory, the facility manager uses a local contractor to perform periodic tests and to respond to check occupant complaints. These laboratories are usually certified and perform water tests according to agreed methods that have been specified by the U.S. Environmental Protection Agency (EPA) and the State or County Health Department. Most labs have excellent reputations and have extensive experience with the utilities. The lab will know what common items should be tested and how often. For an existing facility, after an ongoing positive relationship has been established, the lab will have previous experience and the facility manager can rely upon the laboratory advice.

Most facilities will want to have a working relationship with a laboratory even if they are not required to provide tests. A facility would want to spot check the supply once or twice a year to confirm water is satisfactory. More often these days, tests are required by local regulation. The manager can find these laboratories in the yellow pages of the telephone directory under water treatment. Most labs test water throughout the region and already know the most likely contaminants for which to test. The lab manager recommends the tests.

There are a great many chemicals to test for and the facility manager wants to be careful to test for only those that he needs. A full set of lab tests can cost up to $6,000 but most of the common tests can be provided by collecting a few sample bottles and sending them to the lab. These tests can cost $150-$300.

The best method of dealing with a lab is by letter contract. The lab will usually have one handy and will fax it to the manager. The basic tests take 24-48 hours to run. More complex tests

> ## Case Study: Environmental Spill
>
> Here is on example of how a simple little spill of material can become a complex situation. At a large construction site, a 40-ton crane was being used to move forms that were being assembled on the ground, then picked up with the crane and 'walked" to where the forms could be erected for placing the walls, The assembled forms were large, some 40 ft. square. A gust of wind caught one of the forms as it was being carried by the crane to the site of the next concrete pour. The operator was not able to set the form down quickly and the wind tipped the crane over. Fortunately, nobody was injured but the hydraulic fluid from the crane leaked onto the ground around the crane. After the crane was righted, a large dark brown spot was noticed by personnel where the fluid had leaked into the dirt. The dirt was scooped and placed into barrels and tested to confirm it was "not hazardous." The cost to test the barrels was about $1,000 per barrel for three barrels. The material was confirmed "not hazardous" and taken to a landfill. Had the material not been placed in barrels and had an environmental regulator seen and tested the dirt with the oil in it, the company could have been fined up to $25,000 per day for each day the oil lay in the ground. As the incident was reported to the insurance company concerning the possible damage to the crane, the regulator would have been able to determine the exact date of the spill.

can take a few weeks.

Most laboratories will train the facility staff how to capture the water samples and a cooperative agreement between the lab manager and the facility managers is an excellent way to verify that the facility gets the right tests done in the right way. (For a more-elaborate discussion of water tests and lab services, see Chapter 4.)

Table 1-3. Common water system complaints and most probable causes.

Problem	Probable Cause(s)
Water pressure is low.	Excess demand, pumps off.
Water is too cold.	Water cooler is out of adjustment.
Water is rusty.	Iron pipes, tanks.
Water is cloudy.	Aerators on faucets. Clay in water.
Water is too hot.	Thermostat is set too high.
Too much water pressure.	Pumps too close to faucet or elevation too low.
Water tastes salty.	Vague. Take sample and analyze. Recommend no drinking. Bathing OK.
Worms/Bugs in water.	Shut down system. Treatment has failed.
Dirt, sediment, debris.	Shut down system. Treatment has failed.

Chapter 2

Water:
Supply and Disposal

Where does a facility's water come from? The river? Springs? A well? Conversely, where does water go? Back to the river? Into ponds or to a lake? Because water is a matter of community responsibility, the facility manager should have a basic understanding of water sources.

WATER RESOURCES

People will do whatever is necessary to preserve their water source. People also get very angry if the integrity of their source of water is in doubt. A very small amount of water is consumed—most of it just passes through the facility. The chemistry of the water changes when the water is used.

The Hydrologic Cycle

Each day, the sun warms the sea and part of the water evaporates into the air as steam. Just because the steam is not hot does not mean the water is not there. This water in the air is called the humidity. For any temperature, only a fixed amount of water can evaporate into the air and no more. When too much water is in the air, the steam condenses and falls out in the form of rain.

Because the water evaporated by the sun does not pick up the salt in sea water, rain is fresh and, as it falls on the land, collects into lakes, rivers and streams. Some of the water evaporates when it comes into contact with the earth, some is consumed by plants, and if there is any remaining, it seeps into the ground and continues downward until underlying rock traps it so that it becomes a pool of deep fresh underground water.

When terrain changes and the water from underground is able to drain back out onto the surface, it is called a spring. Because the water was distilled from the sea, fell to earth and was filtered through the earth, this spring water is often very clean and aside from minerals dissolved into the water from the surrounding rock spring water, is usually quite pure. If the spring has been there for a long time, any easily dissolvable minerals have long been washed away.

Eventually, water returns to the sea by flowing overland through rivers, underground, or by evaporating and falling as rain back upon the ocean. Scientists know approximately the total amount of water on the planet and have attempted to track the amount in the atmosphere or on land at any time. Compared to the total amount of water in the oceans, there is very little fresh water on land. But the amount of water used for drinking and human consumption is a very tiny amount compared to the total supply of fresh water.

Most fresh water is used for agriculture. Industry uses large amounts for washing and cooling. Cities use some of the water for bathing and washing and a small amount is used for drinking and food preparation. Only a very tiny fraction of the earth's total water is used as drinking water.

SUPPLIES

City and facility water supplies are available from several types of sources. Nobody makes his own water, except astronauts in space. Because only a small amount of the total water is needed for washing, drinking and bathing, there is enough water to meet the needs of everyone on the planet. However, there are places where there is not enough water. Pollution has also begun to affect the total fresh water supply.

Natural Lakes and River Supplies

Most major metropolitan areas obtain fresh water from lakes or rivers. Water is pumped from the fresh water body. Then it is treated, disinfected and distributed through pipes to storage tanks. From there it is distributed to individual homes, to businesses, campuses, factories, ball parks, etc. Since the source of water is a surface supply, treatment methods must remove debris, organic matter, microorganisms, and some chemicals. If someone upstream has contaminated the river or the lake, the water utility must treat the water to remove the contamination before distributing it to the users.

Withdrawal from natural lakes and rivers usually does not exceed supply—in the West, however, laws allow complete diversion of the entire stream or waterway. Total diversion has been challenged by environmental groups who claim this action is adverse to the environment. Evidence is reviewed in court who decides how much will be left behind, if any. The issue has never been resolved to anyone's complete satisfaction.

Man-made Reservoirs

In order to store more water, man-made lakes have been constructed. Man-made lakes are called reservoirs. Reservoirs are constructed by building a perimeter of earth dikes or embankments and filling it will water from streams, springs or wells. In addition, dams have also been constructed across a river or natural waterway and the gates are closed, backing up the water behind the dam. Municipal water pipelines feed supply water from these lakes.

Reservoirs are carefully managed. Decisions are made each day about how much water is comes in, how much goes out and how much will be stored.

Reservoirs use an area-capacity table as a management tool. The table indicates the volume of water for each elevation of the water line. The facility managing the reservoir knows at any time how much water is being stored by comparing the water level in the lake to the volume shown in the area capacity table. Unless the reservoir is very small, man-made reservoirs also provide recreation opportunities such as boating and fishing. Since water is treated before distributing it to users, this has proven to be an acceptable practice.

Reservoir water must be treated for the same pollutants as the contaminants in natural rivers and lakes.

Springs

Springs are a wonderful natural phenomena, and for hundreds or years, cities were built near springs because the water there was reliable, plentiful and clean. The spring could not be contaminated and it was the central gathering place for the city's most important citizens. Today, few springs are located in a good place for a city as there is not enough access for all of the people who need to use it. Springs are an excellent source of supply if there is enough water.

Another problem with springs is that they occasionally dry up. Spring water, if it has been tested and the components are known, may not have to be treated. However, past animal grazing practices have led to the contamination of many springs. For some utility companies, spring water is captured right at the source and is immediately fed into the distribution system. For other cities, the water is disinfected. Springs sometimes have objectionable odor or color, the result of minerals dissolving in the water.

Wells

Wells are drilled to tap underground water supplies. A well is constructed using drilling equipment to bore a hole through the soil and surrounding rock and into water bearing layers deep underground. Not all wells will produce water. Some do not produce enough water, or the quality of the water is not adequate for drinking. Geologists and scientists use maps, and study rock and geologic formations to determine if there is water present. The presence of other producing wells nearby is a good sign.

The drilling of a well is subject to regulation because the water-bearing layer below the surface may have already been tapped by other users. Drawing water from the ground at one well has the potential to affect other wells bored into the same water table. Figure 2-1 shows a typical well and drawdown curve.

Artesian Wells

An artesian well is one where the natural pressure of the water forces it up through the hole without having to pump it. Few artesian wells remain any longer since the growth of the surrounding communities has put such demands on the well that they now have to be pumped.

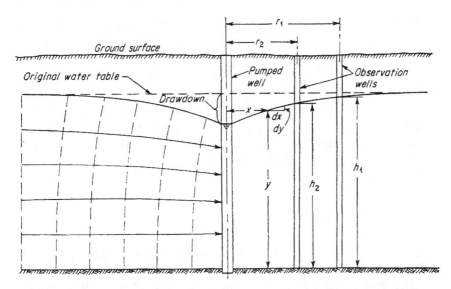

Figure 2-1. Well drawdown. Reprinted from *Hydrology for Engineers* by Ray K. Linsley et al. with permission from McGraw-Hill Book Co., New York City.

After a well is drilled, pumping determines the amount of water the well produces and measures what is called the drawdown. Drawdown is the depth the water table falls while the well is being pumped.

Wells have capacities from 2-3 gallons per minute to several thousand gallons per minute. Pumped wells require electrical energy to draw the water from the ground. The cost of the electricity is factored into the costs of providing the water. The deeper the well is to the water table, the more expensive the pumps and the energy costs are to draw water to the surface. Over time, the water table can be depleted—that is, the water is being drawn out faster than it can run back in. As the depth to the water table increases, the energy required to draw it to the surface increases, raising the cost to provide the water.

Shallow wells, less than 100 ft. deep, are recharged by the immediate surrounding area. Deep wells, 100-300 ft. deep, are recharged over a much larger area. When a deep well has a layer of impervious rock or clay above the water table, it is said to be a protected aquifer. This means that impurities from the surface have difficulty leaching down into the water-bearing layer.

To protect the deep water layer from impurities at higher levels, a well is usually lined with pipes that will prevent contaminated water above from leaking down into the well. The pipe used is called a casing and the wells are called cased wells. Casing a well is expensive because the pipe remains in the ground and becomes part of the total cost of the well. An uncased well, on the other hand, is less costly because the pipe used to bore the well is withdrawn after the drilling is completed.

WASTEWATER

After the facility has used the water, it is wasted. Again, it is rare to completely consume water—usually, it passes through the facility and in the process becomes polluted. Almost all wastewater is treated in some way or another, except for the water from graywater systems. (Wastewater systems are discussed fully Chapter 13.)

Wastewater is carried out of a facility through free-flowing pipes, as opposed the pressurized pipes used to supply fresh water. By designing the wastewater as free-flowing pipe that does not flow full, wastewater is less likely to leak out of the pipes and into the surrounding soils as it would in a pressurized pipe system.

The decomposition of sanitary wastes generates the gases carbon dioxide, hydrogen sulfide, and methane. These sewer gases vent through that portion of the pipe that has no sewer water flowing inside. Provision has to be made to release the gases or else the air pockets will cause the pipes to be plugged.

Sewer mains are buried in the ground below the supply water mains. This reduces the chances of water flowing from the sewer into the fresh water.

At the sewage treatment plant, the natural processes that break down the solid matter in sewage are encouraged. Basic sewer treatment allows the growth of microorganisms that consume the solid matter, breaking it down into inert solids, sludge, and gases.

The wastewater treatment process is upset when chemicals

harmful to the microorganisms (bacteria) enter the treatment process. To prevent this contamination, wastewater treatment plants have a series of tanks, filters and digesters.

If wastewater comes in that contains harmful chemicals, the flow is switched to a backup pond or another treatment tank where the harmful chemicals can be removed before treatment begins.

STORMWATER

The facility manager has to deal with the stormwater that runs onto the parking lot or drains off the building roof during rainfalls or thundershowers.

Provision must be made for adequate flow, or a combination of ponding and flow such that the water does not flood. Stormwater usually channels naturally into gullies, washes or to natural rivers or streams.

Rainfall

The volume of rainfall is calculated from weather data. Stormwater is a function of rainfall intensity, the duration, the vegetation, and the porosity of the soil. For buildings, the rain water runs off. There is a structural calculation that allows for flat roofs to sag slightly and pond some of the water, but this type of design is only for a worst case scenario.

Statistical methods are used by the U.S. Weather Service to predict the storms.

A flood calculated on the basis of a heavy rainfall once in 25 years would be a 25-year storm. For most facility stormwater run-off and storage calculations, a five-year or a 25-year storm is used. The facility manager needs to understand the significance of the storm period.

As the time increases between the assumed heavy storms, the size of the works to carry the water away increases. Culverts and drains for a parking lot or road designed for five-year storms (a storm that happens once every five years) are less expensive than culverts and drains designed for a 25-year storm (a storm that happens once every 25 years).

Parking lots, flat roofs and ball fields must be able to account for most of the major storm expected during the facility life. In addition, culverts and drains must be sized correctly to allow the flow to pass through and prevent flooding.

In practice, engineering designs and codes provide for typical standards acceptable to that area. Engineers that estimate design storms and floods are called hydrologists and can be located in the yellow pages of the telephone directory.

Figure 2-2 shows typical rainfall intensities for estimating stormwater runoff.

Finally, reservoirs and dams are designed on the basis of the "most probable storm." The most probable storm is the estimate of the worst ever case of rainfall. Usually, the estimates for the most probable storm are estimated by scaling up the numbers from a 100-year storm. For major dams and for large rivers, government agencies, usually the U.S. Army Corps of Engineers, calculates floods from the most probable storm. When a flood is predicted,

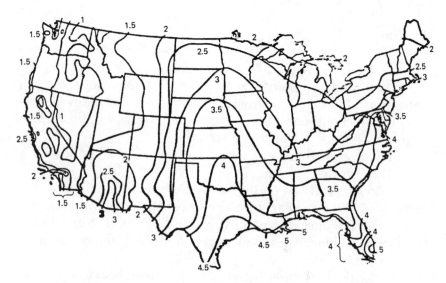

Figure 2-2. Rainfall intensity for the continental United States. The numbers represent total amount of rain, measured in inches, that is expected to fall over a one-hour period every 100 years on average. Courtesy: U.S. Weather Bureau, Technical Papers.

the Corps notifies local government agencies and the media who then notify the public at large. Local offices are listed the telephone directory under United States Government.

Stormwater Ponds

For most facility managers, the problem of stormwater control is one of erosion control. If the campus or property has a tendency to pond, then it is a poor site and steps have to be taken to channel or pump the water to another location.

Because of the tendency of many facilities to be near large cities, there is not any easy location for the runoff water. Hence in recent years ball fields, parks and other grassy recreation areas are recessed slightly to allow rainwater to pond when a large storm dumps rainwater onto the complex.

Erosion Control

For large runoffs, the facility manager must be prepared to deal with erosion of soil on the facility. Erosion caused by stormwater can be halted or arrested. In addition, the construction of underground works to allow the water to pond below the facility is an available option.

Soil Types

Erosion control requires an understanding of the types of soils at a facility. Sand is a granular material. It has no "sticky" material in it to bind up with adjacent grains. Therefore, it will erode quite rapidly when exposed to runoff water. Clay, on the other hand, contains a "binder" that sticks and requires more energy to wash the soil away.

Erosion control is also a function of the volume of water and of the slope of the natural terrain. Erosion occurs when the energy of the water, a combination of its weight and speed, exceeds the resisting energy in the soil.

Methods of Erosion Control

Erosion can be controlled by providing resistant energy to prevent runoff flows from taking the adjacent soil. Large rocks, stones or objects can be used, along with vegetation, rechanneling the water or providing underground pipe to carry the flow off the property. Any combination of these methods can be used.

Vegetation. Vegetation, in the form of grass, shrubs and trees, can be used to hold the soil in place until the worst of the runoff has passed. Landscape architects are the most knowledgeable in providing expert advice for vegetation and planting to reduce erosion from water runoff.

Rock. Large stones placed in the path of erosion will resist the action of the water. Large stone placement is called "rip rap." The size of the stones is a function of the flow of water. For lakes with waves these stones are sometimes two or three feet across. Another method of erosion control is the construction of rock "gabions." A rock gabion is basket of wire mesh or bars containing large stones. Gabions are usually about 4 ft. cubed and contain rounded stones all larger than four inches across. Gabions will last several years and work well for intermittent flows. When the wire mesh rusts away, as it ultimately will, the gabions begin to break down.

Concrete. Concrete has proven its ability to resist erosion in many ways over many years. Concrete can be constructed in retaining walls, sea walls, or simply broken pieces can be placed similar to rip rap. One of the more interesting uses of concrete is in "dolose." Dolose is formed from large blocks and looks very similar to children's jacks, only much larger of course. The advantage of dolose is that it has a tendency to snag other debris in the erosion and helps to anchor them in the pathway.

One old method was the use of abandoned car bodies. These worked fairly well—they were easy to move to where needed and provided more than enough weight to do the job. However, car bodies rust and have oils and lubricants. Old car bodies are not legally allowed and their use would not be recommended today.

Location of Erosion Problems

Keys to locating areas of high runoff are gullies, high water marks, historical recollections and historical records. In recent years, some legal progress has been made at holding upstream managers responsible for concentrating runoff flows. A plowed field would absorb most of the rainwater, but when a facility converted the field to a parking lot, the water ran off and onto the adjacent property. Facility managers must take this responsibility into account when changing the nature of the lay of the land by constructing parking lots or other site developments that are impervious to water runoff. If nothing can be done about the runoff, the facility manager can seek the rights to let the water run off in the form of an "easement." An easement is the right to use somebody else's land for conveyance of runoff water. In some cases, the facility must construct a pipe, underground, that flows clear to a natural stream for runoff flows.

UTILITY INTERFACE

In most areas, the supplying utility tries to work with a facility manager to make sure the overall utility needs are met. These services include assistance in determining the amounts of water supplies needed, in assuring that wastewater facilities are available and accessible, and in making sure the facility is not located in a flood plain or that the site improvements do not create a flooding problem for adjacent users.

Utility support can take the form of measuring flows or estimating total usage. In addition, local government city engineers or building inspectors know what type of rainfall intensity is recommended for their area. These types of details are resolved in facility design and planning processes.

THE NEXT STEP

Water management is the business of deciding what is necessary to supply the right amount and quality of water. In addition, after water has been used, management must decide how it is to be wasted and to make sure it has minimum impacts to others down stream. Facility managers coordinate their activities with local government officials to minimize these impacts and optimize the facilities opportunity.

The next chapter discusses the laws and regulations that affect the facility manager and the facility.

Chapter 3
Law and Regulations

*H*istory has shown that a government that fails to protect its water supply pays a heavy price in disease, sickness, pain and sorrow among its citizens. Because clean water is critical to health, government protects its citizens by enacting laws that control purification, storage, waste and ownership. Understanding these regulations is essential to compliance and avoiding costly penalties.

LAW AND REGULATIONS OF WATER SYSTEMS

In modern countries, the penalties for violation of water law vary from cease-and-desist orders or minor fines to imprisonment. Fortunately, most government regulators have learned that the people are best served if the regulations are applied proactively to prevent water purity violations. For example, if somebody pollutes five miles of river and is caught, does imprisonment restore the river? Facility managers profit from working with the local environmental regulators, provided the regulators demonstrate a level of maturity in administering and enforcing their rules.

It is unfortunate, but in a few cases, government environmental programs have not yet grown from the emotional to the scientific maturity. Regulation is administered with a "now we've got you" mentality.

Government regulation of water starts at the local level, and expands to the state and federal levels. In the United States, federal regulations such as the Safe Drinking Water Act, allow the federal government take swift action if the public health is threatened.

In addition to the federal and state environmental laws, local government has building codes and standards that govern the installation of piping systems that carry and store water.

FEDERAL WATER LAWS

Several federal water laws are summarized here. These regulations are nationally applied and are published in the United States Code. It has been the policy of the United States to allow individual states to manage their own water quality, supplies and pollution programs, provided they are at least as stringent as the federal ones.

Federal water laws are administered by the United States Environmental Protection Agency. Under the law, the individual States assure compliance. Violators can be imprisoned or fined.

Since water laws are updated and amended regularly, a potential facility violation should be referred to legal counsel. Most federal water pollution law is published in Chapter 40 of the Code of Federal Regulations. Maximum fines and penalties are shown in Table 3-1.

Table 3-1. Maximum civil and criminal penalties for environmental violations.

		Action	
Environmental Law	Non-compliance	Willful/Negligent Violation	Withheld/Falsified Information
Safe Drinking Water Act (SWDA)	$25,000/day		
Resource Conservation and Recovery Act (RCRA)	$25,000/day Injunction	$50,000/day 5-year prison	$25,000/day 2-year prison
Comprehensive Environmental Response, Compensation and Liability Act (CERCLA)	$25,000/day	$25,000/day 1-year prison	$25,000/day 1-year prison

The Clean Water Act

Until the late 1960s, clean water was a matter for state control. Individual state regulations managed water pollution, but problems between states and definitions of what really constituted polluted water led to difficulty in agreement and commitment in efforts to clean up polluted waters.

One of the first of a new age of modern water laws was the passage of the Water Pollution Control Act by the United States Congress in 1972. The Act set basic standards against pollution of the nation's rivers and streams. In 1977, Congress revised the Water Pollution Control Act, adding toxic water pollutants to the list and renaming it the Clean Water Act (CWA).

The Clean Water Act continues to be revised and amended with the most recent attempt in 2001. This legislation became bogged down over individual states' ability to implement total maximum daily loads (TDMLs) that would be used to restore pollution-impaired waters. Since complex modeling and sampling is required to establish TDMLs, states lacked the resources to establish them. In the meantime several lawsuits from environmental groups against EPA resulted in court orders directing EPA to establish TDMLs. In FY 2001 Congress directed EPA to utilize the National Academy of Sciences to assist EPA by underpinning pollution control by establishing TDMLs. This controversial work will continue and TDMLs will eventually be used to restore polluted water.

The existing law also provides funds in the form of grants to help communities pay for building wastewater treatment plants. This legislation is called the State Water Pollution Control Revolving Fund or SRF for short. Currently the EPA makes the funds available to the states, which in turn provide the money to communities. In 1996 Congress cut the loan fund significantly. It was partially restored in 1997 and has remained constant at about $1.35 billion per year since 1998.

The National Pollution Discharge Elimination System

Another element of the 1972 Clean Water Act established a permit system requiring facilities to register their pollutant discharges. The intent was to regulate wastewaters and to document several unknowns, at that time, about the releases. Unknowns included the volume of the releases and the constituents of the pollutants in the release. This permission process, still intact, is called the National Pollution Discharge Elimination System (NPDES). NPDES also set limits on certain types of pollutants.

For facility managers, NPDES regulations clarify that discharges into a system served by a publicly owned wastewater treatment plant are not subject to the requirements of an NPDES permit. This means that facilities hooked to public sewer system are not subject to federal NPDES permit requirements. However, the facility is still responsible to prevent pollutants that can have an adverse effect on the sewer plant operation. In addition, discharges from the wastewater treatment plant are regulated by the NPDES requirement.

The U.S. EPA has delegated the responsibility of permit issuance to some of the states, but in others, the U.S. EPA regional offices still manage the permit program because some states have not been willing to take over and run the program on its own.

Regardless of who issues the permits, an application is transmitted from the regulatory agency—either the state or the U.S. EPA to the U.S. Army Corps of Engineers (COE) which examines the application and determines whether the discharge would have an impact on interstate waters. If the COE finds that there is an impact on interstate waters, the COE directs the facility to perform studies to determine the scope of the impacts. The COE then checks the data and confirms the studies.

If this seems like of lot of red tape, it is. But facilities should remember that before this legislation, many streams and lakes were strangling in a sea of pollution and the costs to clean it up so far have been staggering. The effect of the Clean Water Act and NPDES has cleaned up a lot of polluted water and has notified the public of a number of waste discharges that before that time had remained largely unknown.

In September 1995, the city of San Diego was potentially liable for a large fine as a result of unauthorized discharges stemming from a July 1994 waste discharge incident. Figure 3-1 is a copy of a press release published in a trade magazine as a warning to all facility managers.

WARNING—Discharges of wastes from water treatment plants, including chlorinated water, sludge, or chemicals, to surface waters or tributaries, including canyons or storm drains, are in violation of the Federal Clean Water Act unless authorized by a National Pollutant Discharge Elimination System (NPDES) permit. The city of San Diego was potentially liable for penalties (of up to $220,000 as a result of unauthorized discharges that occurred in July 1994.) An NPDES permit should be obtained for all planned or anticipated discharge. Operational procedures should be established to prevent or mitigate unanticipated spills. Staff training and written procedures are critical in elimination of preventable discharges. Utility companies and water suppliers should be proactive in adopting sound operational policies to protect water quality and the environment. This information is provided on behalf of the city of San Diego Water Utilities Department.

Figure 3-1. Case Study: Press release of possible NPDES violation. Courtesy: San Diego Water Utilities Department.

1. Do you have your own source of drinking water? —Yes —No

2. Do you treat the water in your facility? —Yes —No

3. Do you sell water to others? —Yes —No

4. Does your water travel across state lines? —Yes —No

If the answer is "No" to all four questions, then the facility is NOT subject to regulation under the Safe Drinking Water Act.

However, if the facility's water results in an ill occupant, the facility is still liable unless it can be proven the contamination was the not the fault of the facility.

Figure 3-2. Facility manager's test for compliance with the Safe Drinking Water Act.

The Safe Drinking Water Act

Originally passed in 1972, amended in 1986 and in 1996, the Safe Drinking Water Act (SDWA) was designed to protect drinking water supplies. In contrast to the Clean Water Act, which primarily regulates pollutant discharges, the Safe Drinking Water Act regulates water utilities that provide drinking water to users. Drinking water regulations can be complex and confusing because much of the technology used in their administration is based upon laboratory and health-risk science. Facility managers should be aware that the Safe Drinking Water Act does essentially five things:

1)   Any facility that produces, treats, sells or provides water for interstate transport is subject to the regulations of the Safe Drinking Water Act.

2)   The Safe Drinking Water Act sets standards for purity of drinking water (see Chapter 4 and Appendix II).

3)   Facility managers must monitor and sample their water for compliance.

4)   If a facility fails to monitor or monitors and finds impurities in the water in excess of the standards, the public must be notified.

5)   Fines for violations are allowed for up to $25,000 dollars per day of violation.

Originally the law was passed to protect groundwater supplies since the Clean Water Act was, in effect, protecting surface water. But over the past several years the Clean Water Act and the Safe Drinking Water Act have been amended until now, the SDWA regulates primarily suppliers while the CWA regulates primarily polluters.

National Primary Standards

Currently, the U.S. EPA has published the National Primary Standards, which consists of a list of 87 water contaminants. It was the goal of the U.S. EPA to publish another 25 standards every three years, but the U.S. EPA became bogged down in its own rules and regulations, and began to recognize the impossibility of enforcement. In 1996 an amendment to the SDWA relieved EPA of the requirement to publish an additional 25 standards every three years and instead replaced the requirement with a 5-year cycle that requires new standards to be established based upon risks to human health, sound data and science.

A list of the National Primary Drinking Water Standards is included in Appendix II.

Secondary Drinking Water Standards

In addition to the National Primary Standards, the U.S. EPA has published a list of recommended secondary standards regarding taste, odor and color in drinking water. Secondary standards are not enforceable, but are considered recommendations for facilities or utilities or to the individual states. The states have the authority to regulate in excess of the federal standards—therefore, in some states; the recommended secondary standards may be legal requirements.

Monitoring

Facilities are required to monitor drinking water at the tap according to the Safe Drinking Water Act for the pollutants identified in the Primary Standards. The numbers of samples, the times when they are required to be taken, and what each one is to be analyzed for is a function of the size of the system and the number of people served. In addition, since many states are responsible for managing their own programs, a facility manager should contact his state board of health for determining sampling criteria. Chapter 4 provides more information about working with a laboratory and with staffs to determine samples, rotations, and analyses.

Finally, the facility and the facility manager can be fined for not performing the monitoring required by the government body that has jurisdiction at that facility.

Drinking Water State Revolving Fund Program

As a part of the Safe Drinking Water Act amendments in 1996 the U.S. EPA was authorized to establish a State Revolving Loan Fund Program designed to provide funds for improving drinking water quality. Funds can be made available by grants through a state-administered program approved by EPA. Since it is a federal program the process is complex and is coordinated through each state. Additional information on the program can be obtained from the U.S. EPA or the internet at www.epa.gov/safewater.

Public Notification

In addition to the requirement to sample and monitor drinking water, a facility is required to notify the public if the water is not being monitored or if the results of the monitoring reveal impurities in the drinking water in excess of the standards. In general, results must be verified—that is, if sampling reveals water that violates the standards, the facility must sample again as soon a possible and as near to the same source as possible. If these results also exceed the standards, the pollutant is "verified." The facility must then notify the public of the results and of the steps to be taken to reduce individual risks. Notification can be in the newspaper, on the radio or television, by letter or included with the utility bills.

Usually, the facility manger is going to get some help from the regulating agency in a public notification event since the objective is to protect the people's supply. The facility manager should be aware that most of the standards have been set at a level *below* that which will affect humans. Most standards are set at a level that allows an adult to drink two quarts per day for 70 years before there are any ill health effects.

The purpose of the public notice was to make sure the public was aware of the risks to their supply and to allow communities time to take steps to mitigate the impacts before a true health hazard exists.

HAZARDOUS WASTE LAWS

In addition to complying with federal water laws, the facility manager's water management role can be affected by three federal hazardous waste regulations for which the fines and penalties are much greater than in the water regulations. These laws are written to protect the public from illegal dumping of hazardous wastes, but are presented here because the activity of the facility manager in water management can lead to generation of small amounts of hazardous waste.

Resource Conservation and Recovery Act

Originally written in 1976 and revised in 1984 and again in 1991, the Resource Conservation and Recovery Act (RCRA) was written to regulate solid and hazardous wastes. Facility managers who generate, transport or dispose of hazardous wastes must comply with hazardous waste rules. The U.S. EPA has posted a list of materials that are considered hazardous or general criteria for defining hazardous waste.

Hazardous waste management is a relatively new field for facility managers and it is recommended that a consultant be hired to assist the facility in this area. CFR 40 lists most of the hazardous wastes along with the characterization codes.

Finally, the facility manager should be aware that hazardous waste regulations create "cradle to grave" accountability. If the facility hires a contractor to dispose of waste and the contractor is negligent in his duty, regulators can come back to the facility that generated the waste and require the facility to pay additional costs of further disposal.

Comprehensive Environmental Response, Compensation and Liability Act

The flaw in the RCRA law was that it related to generators and processors of hazardous wastes. It did not apply to spills or abandoned hazardous waste sites that already existed. In 1980, the Comprehensive Environmental Response, Compesation and Liability Act (CERCLA) was passed to deal with this area of hazardous waste and cleanups. CERCLA is sometimes nicknamed the "Superfund" because funds were provided under the law for cleanup of the waste sites. For the first few years, CERCLA had difficulty, primarily because most companies found that it was cheaper to fight legally with U.S. EPA than to clean up the spills.

Superfund Amendments and Reauthorization Act

As a result of CERCLA's severe and strict rules making it less costly for the hazardous waste generator to fight the law in court than to try to contribute to clean up the spills, the U.S. Congress revised the CERCLA law in 1986. The revisions were intended to free up more money for cleaning up spills and waste less money on costly legal battles that were not effective in cleaning up spills and hazardous waste sites. The revisions were packaged as the Superfund Amendments and Reauthorization Act (SARA). In addition, SARA created the requirements for informing the public of the presence of hazardous chemicals and for emergency response capability by the facility and the community.

Hazardous Waste Management

Facility water and wastewater managers will be affected probably more by RCRA than by CERCLA or SARA. Under these laws, however, regulations defining hazardous wastes and handling and transporting of them are similar. In general, any facility that generates hazardous wastes must establish a program for handling and managing waste to prevent it from spilling into the environment.

Hazardous wastes are defined as hazardous material *no longer fit for its intended purpose* and either listed or characterized as a hazardous waste by the U.S. EPA. A listed waste is one that is on the U.S. EPA's list. Most facility water managers will not be required to deal with listed hazardous wastes. However, a waste is "characterized" if it exhibits one or more of the following characteristics: ignitable, corrosive, reactive, or toxic (see Table 3-2).

Finally, the facility manager who is a generator of hazardous wastes must maintain records of waste generation and especially shipment. A manager who generates a small quantity of hazardous waste must obtain an U.S. EPA number and prepare a Hazardous Waste Manifest for shipping waste to a treatment center. The intent is to maintain the cradle to grave management of hazardous wastes.

LOCAL IMPLEMENTATION OF FEDERAL STANDARDS

The State

Water supplies are regulated by individual states in most

Table 3-2. Definition of hazardous waste by characteristics.

Ignitable   A waste liquid, solid or gas that will flash and burn. For liquids, a flash point less than 140 degrees F. This class also includes oxidizers such as bottled oxygen.

Corrosive   A waste liquid with a pH lower than 2.0 or greater than 12.5, or a waste liquid or solid that corrodes steel at a rate greater than 1/4-inch per year.

Reactive   Wastes that react when handled improperly or when mixed with water or air. Examples include explosives and propellants.

Toxic   Poisons or chemicals that leach into poisons after the performance of tests known as the Toxic Chemical Leaching Procedures (TCLP).

cases. The state's board of health, working with federal and local agencies, monitors and reports on the health in the given state. A state agency requires the tests of water supplies for pollution and contamination and checks county and municipal supplies for contamination as well.

Operators of many systems are required to check the water on a regular basis. The results of the tests are examined randomly and periodically for compliance with local, state and national standards.

The states work jointly with the federal government to implement the national laws. States can have laws which exceed the Federal Standards but cannot have laws less stringent than the Federal Standards. This means the states can be more restrictive about pollution but cannot be less restrictive. Individual states that perceive a problem not yet recognized by the federal government can act, on their own, to remedy it.

Finally, under the various federal laws—Clean Water Act, Safe Drinking Water Act, RCRA, CERCLA and SARA—a state can run the programs as long as the state operation meets the requirements of the U.S. EPA. In the case of the Safe Drinking Water Act, for example, all of the states and territories have taken over the running of the program. The state spot checks the monitoring requirements, makes sure the waters are safe and notifies and fines non-compliant facilities.

Various other laws, such as the Clean Water Act and RCRA, have been delegated to some states but others are still administered by U.S. EPA. When a state takes over the management of one of these programs, the state has "primacy" over that program. Each of the states, therefore, has primacy over the Safe Drinking Water Act but not all states have primacy over the Clean Water Act and RCRA.

The federal facility manager can determine who has the "hammer" by calling the local U.S. EPA region and asking if the state has "primacy" over the program in question.

The County

Under the jurisdiction of the state, the county or other utility responsible for water and wastewater systems collects the data and reports it to the appropriate state agency. Through the federal and state governments, funds can sometimes be made available for treatment works.

The Municipality

Depending upon the relationships between states, cities and counties, the local government responsible for water or wastewater collects the data from discharges or supplies and forwards it to the states. Through the county, state and federal governments; funds can be made available for testing and for construction of new treatment works.

The U.S. EPA

The U.S. EPA is charged by the U.S. Congress to implement all of the laws discussed in this chapter. The U.S. EPA prepares national standards for drinking water supplies. The U.S. EPA estimates there are a total of 74,000 separate systems serving water to most American citizens. An individual system, such as a well, cistern, or private spring is not included in this list.

The U.S. Army Corps of Engineers

Situated in different regional boundaries from the U.S. EPA, the U.S. Army Corps of Engineers (COE) assesses applications for NPDES permits in most states. The U.S. COE reviews applications and determines the impacts to the state and intrastate waters.

CODES AND STANDARDS

Besides managing the water system itself, facility water managers also must deal with various codes and standards that relate to the management of the systems that provide water to the public. These codes and standards, called building codes, are intended to make sure the public receives an adequate flow of water and that the plumbing and the facility is "safe" for the users. Like the federal laws the code is a law and the facility must comply with it. Usually a model code agency, the proponent of the code, recommends it to the general membership. A city, town, township, state or community then "adopts" the code by resolution and it becomes law.

Unlike the federal water codes, building codes/laws are adopted by ordinance and are more subject to local control.

The agency or proponent of the code is usually comprised of people who are familiar with the industry, who serve with the model code agency through membership, and who have input to the code.

A code proponent body is democratically run—any member of the code group can propose a change and the membership. After review, members vote to decide whether to adopt the change or study it further. (See Chapter 19 for a list of most of the associations that draft model building codes.)

Code enforcement is left to the local jurisdiction. Usually, city or county building inspectors, or sometimes local fire marshals, inspect facilities to make sure the codes or ordinances are met.

National Fire Protection Association

A large organization with several hundred thousand members, the National Fire Protection Association (NFPA) writes model fire protection and building codes that have been adopted in many areas of the United States. The NFPA codes also address water systems since water is used to fight large fires. Comprising

17 volumes, the NFPA codes include fire-resistive construction; exits and openings to allow people to escape burning buildings; lighting; electrical work; fire sprinklers; fire alarm systems; water supplies for fire protection; and a number of other miscellaneous elements such as flammable gas storage, emergency response, fire equipment and safety. The NFPA meets twice annually to discuss and update the codes.

Uniform Building Code

Throughout the United States, several model code agencies have codes for building construction. In the Northeast, the Building Officials Code Administrators (BOCA) publishes the Standard Building Code (SBC). In the South, the Southern Building Code Congress International (SBCCI) publishes the Southern Building Code (SBC), and in the West, the International Conference of Building Officials (ICBO) publishes the Uniform Building Code (UBC). In many areas these codes are similar, defining the criteria and standards for constructing buildings.

The major differences between the codes are regional ones derived mostly from the use of different building conditions and materials. For example, the UBC in the West gives general information about the use of western fir wood, which is not as available in the South or in the Northeast.

In general, all of the codes mentioned in this section relate to the building shell and not to the internal components.

Uniform Plumbing Code

Several code organizations have their own plumbing codes, making plumbing standards regional and discontinuous. One common plumbing code that is extensively in use is the Uniform Plumbing Code (UPC) published by the International Association of Plumbing and Mechanical Officials (IAPMO) of Ontario, California.

The intent of all plumbing codes is to assure that supply is adequate and that drainage is sufficient. As can be seen from the discussion of clean water laws earlier in this chapter and of microorganisms in Chapter 4, poorly managed water can become dangerous to the public. By conforming plumbing systems "to the code," the facility manager reduces his risk because the code (law) is written by professionals who have built similar systems.

Regulated Construction Materials

Most code organizations test and approve materials used for plumbing and piping systems. In general, a manufacturer must submit his materials to the code association for tests to confirm the item or material meets the standards of the code. Code groups publish lists of vendors whose equipment or materials meet the tests.

These devices are "listed" and it is often specified in contracts that only listed items are allowed in the construction. If construction is required to be subject to the code, and most new construction is subject to the codes, then the codes themselves require the materials be "listed."

Other Code-making Bodies

Just as there are associations for the fire, building and plumbing codes, many other associations and trade groups provide input to codes. Professional organizations continue to develop standards for cost effective and safe systems. Some of these professional associations are the American Society of Civil Engineers, the National Sanitation Foundation, The American Society of Mechanical Engineers and the Plastic Pipe and Fittings Association (see Chapter 19).

WATER RIGHTS

Just as there are laws regulating pollution, water purity and plumbing and piping, water is property and there are rights to use water as well as rules against pollution. In many states, legal title to water is the same as legal rights to coal or gold.

The two main types of water rights are adjudicated rights and riparian rights.

Adjudicated Rights

In the Western United States, which is mostly and desert, water rights are governed by adjudicated rights. In the days of early western settlement, a farmer or rancher quickly found there was not enough natural rainfall to keep the crops alive until harvest. The crops would sprout and take root in the spring, but with the approach of warm summer days the crops would wither and die before harvest. To keep the crops growing, the farmer irrigated the fields by channeling water from a stream or river onto the fields. The channeled flows would sometimes keep the crops alive until the harvest could be gathered in the fall.

As more farmers moved onto the land, however, ditches were cut into the river upstream from the original farmer. When the original farmer wondered where the water in the river had gone, he went upstream only to find another farmer had diverted the stream and was using the water instead. The farmer faced with certain starvation if he did not get the water onto his own crops would destroy the upstream farmer's irrigation system. Since the upstream farmer was faced with a similar situation, both farmers were forced to settle the issue in the Old West fashion... with six-guns. Since this method of settling arguments was not good for the local community, the concept of an adjudication of the water was born.

In an adjudicated right, the first farmer to put water to beneficial use is allotted a portion of the water. Thus the first right became his. The next farmer receives the next portion and so on down the river until all the water in the river is accounted for. Today, each owner is allotted so many gallons based upon complex formulas of water application. The state agency responsible for administering these rights is tasked with making sure that each farmer or owner receives his share from the total allocation.

Under this concept, the rights to the water were combined with the rights to the land. Later, when large canals and complex water transportation systems made moving the water more feasible, the water was transferred separately from the land. For example, the land could be sold, but the water rights retained, or the water rights sold and the land retained. Many a western water swindle occurred when a potential landowner was shown land adjacent to the river with a large canal right next to his property, only to discover that his newly purchased land rights did not include the water rights. Later, this type of dishonesty declined when local banks began to make loans for land deals.

Riparian Rights

Contrary to adjudicated water property rights common in the arid western states, riparian rights, common to eastern states, are conveyed to landowners adjacent to rivers lakes and streams. The basis for riparian rights is that there is essentially enough water available for everybody and the adjacent owner to the river or lake may take water freely from the adjacent water body as long as significant changes to flows or levels do not affect others.

Consumptive and Non-consumptive Use

There are two types of use patterns discussed in the management of water rights. These are consumptive use and non-consumptive use. For most facilities, use is of the non-consumptive type. The water comes into the facility and is used for washing, flushing, rinsing, etc. and then it is returned.

When returned, it is more polluted, but essentially the same volume is returned.

Consumptive use means that the facility owner consumes the water on his property. Mostly consumptive use occurs in power plants where it is used for cooling and it is evaporated into the air as it cools the power turbines or other equipment.

Chapter 4
Water Impurities

*P*eople are getting sick in the facility, and the cause is traced to the water. Lost productivity and a lawsuit ensues. Then the building owner has to tell the public. How could this have been prevented? The only way to know if water is safe is to have it tested regularly. Understanding the dangers of impurities in water, including the various types of problems that occur, is essential to properly overseeing water testing and purification.

THE ROOTS OF WATER PURITY

People talk about pure water and wonder if their water is absolutely safe. In fact, no water is completely pure and the term "absolutely safe" depends upon so many variables that industry professionals qualify the term with relative factors that will be discussed here. A quick look at Figure 4-1 and a short discussion of water chemistry will help clarify this issue. The figure shows a sketch of the water molecule which is made up from two atoms of hydrogen bonded to one atom of oxygen.

The oxygen atom has eight protons (positive charge) and six electrons (negative). It is the imbalance of the electrical charges that causes oxygen to want to bond, which is why pure oxygen is such a reactive gas—it is a key element of fire, for example.

The hydrogen atom has one proton and one electron. The electron on the hydrogen atom electrically bonds in the two free positions in the oxygen atom and forms a water molecule.

It can be seen from the figure that one side of the molecule has negative attractions and the other side has positive attractions. Because of the shape of the molecules, other chemicals dissolve in the water. This is why it is difficult to have water that is absolutely

43

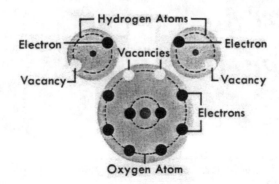

A water molecule consists of two hydrogen atoms and one oxygen atom. Each hydrogen atom has room for another electron around its nucleus. The oxygen atom has room for two more electrons.

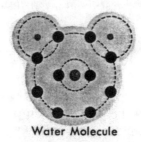

Water Molecule

The two hydrogen atoms and one oxygen atom fill their empty spaces by sharing electrons. The resulting water molecule Is an extremely tight structure because Its atoms share their electrons.

Figure 4-1. The water molecule.

pure. Dirt or soil will soften and dissolve in the water along with salts, acids and many other components.

The ability of water to dissolve many chemicals accounts for ionized particles in water that could be called impurities. These particles include mineral salts, metals, chlorides, acids and others.

As a result, the purity or "safety" of water is relative. The only way to know if the water is safe is to have it tested. A complete battery of tests can cost as much as $3,000. Depending upon the size of the facility, several samples may be required from different areas at different times of the year or season. Basic tests are less expensive and can be performed for as little as $200.

Completely pure water is "hungry" and wants to bond with other minerals. Extremely pure water is difficult to attain and even harder to maintain because of the aggressive nature of the molecules. In general, however, this desire of water to bond with other minerals is what makes it such a valuable substance since cleaning, washing and all life depends upon these chemical bonds.

As with all molecular chemistry, a molecule has different characteristics from its constituent elements. Oxygen is a reactive gas, as is hydrogen. When they react together, the hydrogen burns in the oxygen, releasing a lot of heat that results from molecular bonding. The result of the reaction is water, but the heat from the reaction is so strong that the water forms steam. If the reaction is very carefully controlled, the water from the reaction can be captured.

WATER IMPURITIES

Impurities in water are relative. For example, impurities in a wastewater may be a chemical that poisons the bacteria that will purify the water over time (see Chapter 11, where we discuss water purification techniques). In drinking water, impurities can make people ill; other impurities give the water flavor. For a few types of water supplies, the absence of mineral impurities will cause the water to have a bland taste. In other water, the presence of mineral salts are thought to enhance health.

Water impurities can take one of two basic forms—either the impure particles are suspended in the water, or the elements of the impurities have separated from their basic form and have dissolved in the water. Suspended particles can be removed by filtration. A dissolved impurity is more difficult and costly to remove than a suspended impurity. Figure 4-2 shows some relative impurities in water and the sizes of the impurities. Impurities can be living microorganisms or they can be mineral or organic chemicals.

Examples of impurities in drinking water include turbidity, dissolved salts, dissolved metals, microorganisms, living organisms, radionuclides, volatile organic compounds and pesticides.

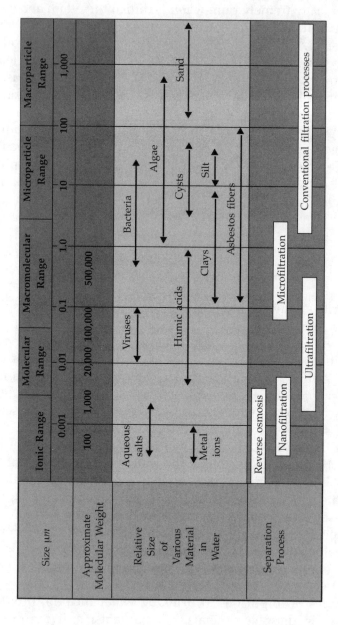

Figure 4-2. Typical water impurities and selected separation processes. Reprinted from *Journal*, by permission, ©1995, American Water Works Association.

Nationwide, a committee tracks and reviews current and pending regulations concerning drinking water supplies. The document, called Drinking Water Standards, is available from the **Office of Ground and Drinking Water (4601), Ariel Rios Building, 1200 Pennsylvania Ave., NW, Washington, DC 20460-0003; 1-800-426-4791.**

Action Level of Contaminants

Each impurity has a recommended standard. It is important that the facility manager understand the basis for level set by the regulations. The U.S. EPA has set the action level *well below* the amount where there are known ill health effects. The purpose is to require facility managers to notify occupants of an impurity in the water supply in time for the public to take action.

While a normal person's health will not be at risk if the standard is exceeded, there is the potential that elderly, infirm, children or individuals with contributing diseases may experience a degree of increased health risk.

Turbidity

A common, easily recognized water impurity is turbidity. Turbidity is the presence of free suspended particles in a water supply. Because of their small size, the particles will not readily settle out, but given enough time, these particles settle to the bottom of the basin. The particles obscure light from shining through the water. In water analysis, turbidity is used to signal the presence of other chemicals or plant life that might have potential ill health effects. Hence, the general application: if water is not clear, do not use it.

Turbidity can simply be an inorganic particle of very small size that is evenly dispersed in the water causing a cloudy appearance that poses no ill health effects. Turbidity is measured in Nephelometric Turbidity Units (NTU). An exact reading is obtained from a laboratory, but many labs will make a color sheet available. By comparing a sample of water from a specific source to the color sheet, an estimate of the turbidity can be made. By itself, turbidity in drinking water can be insignificant, provided the source of the turbidity is known. For example, a fine clay, evenly distributed in well water, can be completely inert and have no effect on the user. Since turbidity can block a lab analysis for other chemicals and microorganisms, however, it is regulated in water supplies.

Turbidity can include both inorganic and organic particles, but it does not include dissolved salts, which will not settle out and must be removed by an alternate technology.

Dissolved Salts

As previously discussed, common dissolved salts in water include sodium chloride, sodium bicarbonate, calcium carbonate, calcium sulfite, magnesium carbonate, magnesium chloride and other salts derived from magnesium, calcium, potassium or sodium.

Water Hardness

Throughout the country, water supplies derived from wells contain dissolved mineral salts. These salts from either calcium or magnesium are combined with a carbonate ion in the forms, $CaCO_3$ or $MgCO_3$.

Salts leave a film residue when evaporated and either calcium or magnesium remain on the sides of pots, pans and bathtubs when water evaporates. The white film is a salt called calcite which is actually calcium carbonate. Calcite when left as residue on the side of a vessel is hard to dissolve. Calcite, by the way, is the same mineral that makes stalactites and stalagmites found in caves. Because the presence of these salts make it difficult for soap to form a lather, the water with calcium carbonate in it is called hard water.

Both calcium carbonate and magnesium carbonate will dissolve in a slightly acidic compound such as citric or muriatic acid. The hardness component of water is measured with a test kit that measures the amount of calcium carbonate dissolved in water.

The calcium ions can be measured in unit weight per gallon and are usually referred to in grains or grains per gallon. (One grain per gallon is equal to 17.12 parts per million.) A complete chemistry analysis will indicate hardness or the presence of both the calcium and magnesium ions.

Water with less than 100 ppm of calcium or sodium is considered soft water. Water in excess of 220 ppm is considered hard water. Most hard waters come from wells in the Midwest region. Water supplies that are primarily derived from surface water are not as hard as well water.

The use of grains per gallon (gpg) is used in sizing water softeners which remove calcium and magnesium ions. The relationship of the number of grains of hardness and the amount of water to be softened is used by water softener manufacturers to calculate the amount of water softener resin needed (see Chapter 11).

Hard water has no ill health effects up to 300-400 mg/liter concentrations.

Other Salts

Sodium chloride, or table salt, is sometimes present in water supplies but does not show up as hardness in a water hardness test. Too much sodium has the potential to affect individuals with cardiovascular disease; however, most sodium received by individuals is in the form of table salt added to foods rather than sodium dissolved in a drinking water supply.

Sulfates and sulfites in water supplies will give water an unpleasant odor. High concentrations of sulfates can create stomach upsets and will have a laxative effect on the lower bowel. Table 4-1 lists some common salts found in water supplies.

Table 4-1. Dissolved salts in water supplies.

Sodium chloride	NaCL
Magnesium chloride	MgCL
Calcium carbonate	CaCO3
Magnesium carbonate	MgCO3
Calcium bicarbonate	Ca2(CO3)2
Magnesium bicarbonate	Mg2(CO3)2
Calcium sulfate	Ca(SO4)
Magnesium sulfate	Mg(SO4)

Dissolved Metals

In drinking water supplies, the presence of dissolved metals potentially poses a problem to the facility manager. Of the most significance are lead and copper. In general, these metals do not normally dissolve in water supplies unless the water carries a slightly acidic component whereby lead or copper from the pipe materials dissolves in the water.

Copper

Copper enters water piping through acidic decomposition of the water piping. In rare other cases, it can be present in well water supply or in surface water supplies from copper bearing minerals. Copper in small concentrations can pose ill health effects in the form of stomach or bowel discomfort but effects are not permanent.

Table 4-2. Definitions of salt concentrations.

"Saline"	<42,000 ppm
"Slightly brackish"	1,000-3,000 ppm
"Brackish"	3,000-10,000 ppm
"Sea water"	32,000-36,000 ppm
"Brine"	>42,000 ppm

Lead

Similar to copper, lead enters water supplies through acidic decomposition of water piping, particularly in old buildings where lead was used as the piping material. In addition, many copper and cast iron pipe systems were joined with lead or solder with a high lead content. The lead elements break down in mild acidic water and mix with the drinking water. Lead concentrates in the body and its effects are cumulative. Lead is a known cancer risk, has adverse kidney and nervous system effects and is especially toxic to infants. Unfortunately, studies for lead contamination by the U.S. EPA in the late 1980s revealed that up to 20 percent of the nation's water was contaminated with lead.

Elimination of lead and copper can be accomplished by treatment, usually by adding chemicals to the water to reduce the acidity to deter the corrosive action.

Other Metals

Zinc, aluminum and iron are not currently regulated but have recommended maximum levels. Health effects are minimal but they can give water metallic taste or unfavorable color.

Mercury, once thought to be inert because it does not mix with water, was discovered concentrated in shellfish on the edges of large industrial areas. Mercury causes nervous disorders.

Inorganic Compounds

Metals such as antimony, nickel and beryllium are covered in the standards under the inorganic compounds because their presence can cause kidney or liver effects. In addition, nitrite and nitrates are regulated because these minerals have been shown to affect the blood/oxygen cycle in infants. This problem is called "blue baby syndrome." Finally, asbestos is regulated under inorganic compounds because it has been shown be a cancer risk for lung tumors.

Radionuclides

Trace elements of radon, radium, uranium and other radionuclides have been discovered in some water systems. These traces are very small but are still detectable by sophisticated laboratory techniques. All are suspected to increase cancer risk. Radionuclides generate ionized radiation. That is they emit radioactivity according to their unique atomic structure. The health problem of ionizing radiation is that it interferes with cell replication and regeneration, causing the cells to mutate. An uncontrolled growth of mutated cells manifests itself as a cancer.

Radioactivity is emitted from radionuclides in three forms, Alpha particles, Beta particles and finally Gamma rays. The Alpha particle is a very small particle consisting of two protons and two neutrons. Since the particle has no encircling electrons, it wants to take them away from atoms that have electrons. Beta particles are free electrons, and since they have no proton to balance their charge, seek to attach to atoms with extra protons. Gamma rays are high energy rays that affect the alignment of molecules. All three affect living cellular tissue.

In drinking water, the presence of Alpha Particles and Beta Particles and Proton emitters is regulated, as well as Radium and Uranium.

Organic Compounds

Another category of water impurity that affects drinking water are chemicals commonly referred to as organic compounds. Organic compounds include natural organic chemicals, derived from petroleum products like gasoline and kerosene, and man-made organic materials like pesticides. Many of these types of products are regulated and are specifically addressed in the standards. The presence of these chemicals in water pose nervous system, kidney and liver defects or cause cancer risks. Limits for these chemicals in drinking water are low, in the order of 5-100 parts per billion (ppb).

Sulfur Made Simple

Sulfur items are sometimes confused. Sulfide is the combination of a pure element sulfur. H2S is hydrogen sulfide. Sulfite is the ion SO3. Copper sulfite is $CuSO_3$ and hydrogen sulfate is H2SO4, commonly known as sulfuric acid.

S	Sulfide
SO3	Sulfite
SO4	Sulfate

In water supplies, pure sulfur will combine with hydrogen ions and oxygen ions, forming hydrogen sulfide, H2S. Hydrogen sulfide is responsible for the "rotten egg" smell in low concentrations and is, by itself, a poisonous gas in higher concentrations.

Sulfites and sulfates in water give an odor and taste which is unpleasant in high concentrations. High sulfate concentrations have been known to cause stomach cramps and diarrhea.

MICROORGANISMS

When it comes to pure water, the presence of microorganisms in drinking water supplies is a heavily discussed topic. Who has not had a friend go Latin America and come back with horror stories of massive stomach cramps and diarrhea? Of course, microorganisms (bugs) get blamed for these horror tales. Microorganisms in water account for several outbreaks of disease in this country as well. In 1993, a *Cryptosporidium* outbreak in Milwaukee's water supply sent thousands of people to the hospital. *Legionella,* another microorganism, is responsible for Legionnaire's disease, a disease that results from *Legionella* bacteria growth in cooling water supplies that are picked up by air conditioning systems.

Living Impurities

In a drinking water supply, the presence of cells indicates the tolerance in the water supply for living microorganisms.

Cells, which are extremely large compared to water molecules, live in all water systems. The membrane of the cell wall is tough enough to withstand the dissolving action of the water molecule. But chlorine, which has been added to water supplies for the past 150 years or so, reacts with the cell wall. In a simplified explanation, all a chlorine molecule has to do is touch a cell and the tough outer membrane of the cell is broken. Then the water molecules can break through into the cell and destroy it.

Not all cells in water are harmful, and in the case of wastewater treatment, a certain type of bacteria is used to break down the impurities and ultimately purify the water. In general, all cells are referred to here as microorganisms. Cells that cause disease and sickness are call pathogens.

Drinking water purification experts and the industry have long used an "indicator" organism to aid in the test for water impurities. This is because the test for a true germ that poses a health risk, if positive, would mean it was already too late for the public official to take any action. For example, if *Vibrio cholerae* (the microorganism that causes cholera) were found in water supply samples taken at the tap, a health warning would arrive too late since the contaminated water would have already been used by some number of the population. In the case of pathogens, the worst risk comes from fecal coliforms contaminating a drinking water supply.

E. Coli

E. Coli, short for Escherichia coli, is used as a measure of the potential harmful bacteria in water supplies. The E. coli bacteria is chosen as the indicator because it is consistently present in human feces in large numbers, and it has the same survivability in water as more pathogenic organisms. The test for E. coli is fairly simple and a simplified diagram for its test is shown in Figure 4-3. Another common test method uses an enzyme that changes color if the coliforms are present. If the coliforms are of the E. coli variety, the enzyme color fluoresces. Kits of this type are available from specialized laboratories. Follow all instructions and necessary safety precautions.

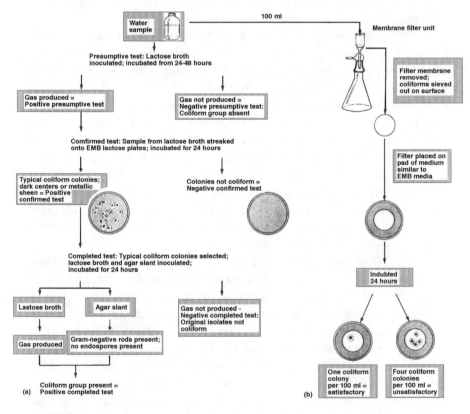

Figure 4-3. Specialized method for sampling for coliform bacteria. Courtesy: The Benjamin/Cummings Publishing Company, 1992.

In the past, there have been coliforms found as a result of growth on the inside of water supply piping. These coliforms, while not the result of human consumption, have led to some boil water orders in community supplies. Called *biofilm*, these coliforms have not been confirmed to be hazardous to public health. It is important to separate the presence of total coliforms from *E. coli* and to have data on coliforms resulting from growth inside water supply piping.

Chlorine is the main chemical used to treat water supplies and to eliminate coliforms. Since the pattern of coliforms is similar to other cells that are true disease-causing germs (pathogens), chlorine neutralization is effective in removing the pathogens as well.

Viruses

Similar to coliforms, viruses are carried in the wastes of humans and animals; however, they are neutralized by chlorine in much the same way as coliforms. Therefore, the tests for *E. coli* are used as an indicator for the removal of viruses and most other pathogens.

Giardia

Increasingly difficult to detect using *E. coli* as an indicator organism and more difficult to kill than coliforms, the cysts of the *Giardia lamblia* grow into a tiny one-celled animal that uses a sucker to attach itself to the intestinal cell wall. Persons infected with microorganism have the disease giardiasis, a prolonged diarrhea-causing disease that results in nausea, weakness, weight loss and abdominal cramps. A picture of the grown protozoa is shown in Figure 8-4. About seven percent of the total population carries the microorganism and sheds the cysts in their feces.

The laboratory procedure for *Giardia lamblia* is a complex fluorescent antibody test.

The *Giardia* cysts are extremely resistant to chlorine and hence another method of treatment is recommended. Most often, this method is filtration of the water supplies because the large cyst is filtered out under controlled filtration methods.

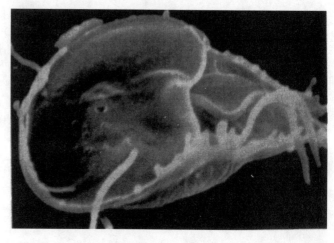

Figure 4-4. Electron microscope photograph of a full-grown *Giardia lamblia*. Courtesy: Visuals Unlimited, New Hampshire. Reprinted from *Microbiology: An Introduction* with permission of Benjamin/Cummings Publishing Co., Inc.

Cryptosporidium

Like *Giardia*, *Cryptosporidium* is shed in the feces of humans and animals. In the environment this microorganism forms a protective cap to protect itself. The encased protozoa is called an oocyst (pronounced oh oh cyst.) A photograph of the oocysts is shown in Figure 4-5. Like other protozoa, only one type of the animal is known to infect humans. This type is called *Cryptosporidium parvum*. *C. parvum* are active in surface waters, and in shallow wells subject to contamination from surface waters.

Individuals infected with *C. parvum* have the disease cryptosporidiosis which is characterized by nausea, vomiting, fever, diarrhea, headache and loss of appetite. This disease often passes within two weeks or less and most normal individuals recover fully in that time.

At the present time, surface waters must be filtered to remove *Cryptosporidium*. The current test for *C. Parvum* is the fluorescent antibody test similar to the one for *G. lambia*.

Like the *Giardia* cysts, the *Cryptosporidium* oocysts are extremely difficult to kill using chlorine. The size of both the *Giardia* and crypto cysts exceeds one micrometer and they are removed by filtering water below that size.

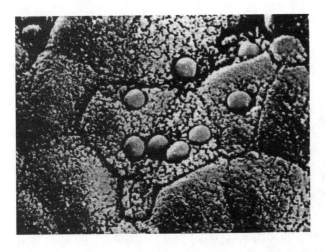

Figure 4-5. Electron micro- scope photo- graph showing the oocysts of the *Cryptosporidium parvum* **microor- ganism. Cour- tesy: Diagnostic Medical Parasitology, L.S. Garcia and D.A. Bruckner, ©American Society of Microbiologists.**

Legionella

Unlike the *Giardia* and *Cryptosporidium* which form cysts in raw and treated water, the *Legionella pneumophilia* bacteria infects individuals by airborne contamination along with ingestion through drinking water. The bacteria cause Legionnaires' disease, a severe pneumonia. The source is usually not the drinking wa- ter supply but rather the humidity in the air conditioning sys- tem.

L. pneumophilia grows in cooling towers in the presence of dirt and mud but the bacteria is unique because it grows in tepid water instead of the usual supply of cold water. The bacteria is picked up in the air and carried to individuals along with the tiny water droplets. Evidence to date is not clear whether the bacteria are eradicated by chlorine, but facility managers should take the precaution of treating water-heating and cooling-system waters with prepared biocides according to manufacturers' recommenda- tions and using all necessary safety precautions. In some cases, the application of hypochlorite (liquid bleach) will adequately kill populations of *L. pneumophilia*. The newspaper article in Figure 4- 6 indicates what can happen to facilities that fail to manage their water supply and do not test for the presence of *Legionella*.

For more information about treating heating, ventilation and air conditioning systems for *Legionella* and other airborne bacteria, see *Indoor Air Quality: A Guide for Facility Managers* by Ed Bas, a companion volume in The Facilities Management Library.

Deadly Bacteria Resurface in Offices

By Lisa Daniel
Federal Times Staff Writer

The bacteria that causes Legionnaires' disease has returned to a California Social Security Administration building. Two people there died from the disease in 1991.

Small amounts of the *Legionella pneumophila* bacteria were found April 11 in the water system at the Social Security Western Program Center in Richmond, Calif., SSA officials said.

The bacteria was found during the first test of the building since officials reduced such checks from monthly to quarterly, said Howard Egerman, health and safety representative for the American Federation of Government Employees Council 220.

The agency conducted monthly checks for *Legionella* since September 1991 when the building was evacuated because of an outbreak of the bacteria. Two women who worked as janitors died from the outbreak and 13 people became sick from it.

The agency will return to monthly tests for the bacteria and is injecting small amounts of chlorine in water in the Richmond office, said Pam Reim, a spokeswoman in the agency's regional office in San Francisco.

"We're doing this rigorous testing that, for the most part, other buildings don't do," Reim said.

Bacteria recently were found in three areas: two sinks in the fifth floor men's and women's restrooms and a basement janitorial closet sink where it was found in 1991.

The latest outbreak was much smaller than that found in 1991 and did not put employees in serious danger, Egerman said.

"This was unfortunate, but it could have been a heck of a lot worse," he said.

(Continued)

Figure 4-6. Case Study: Legionnaires' disease strikes an office building. Courtesy: ©1995, *The Federal Times*.

Still, the incident sent shock waves through employees who remember the 1991 outbreak.

"A lot of them have never forgotten what happened," Egerman said.

Several people took the day off after the latest incident and there was at least one request for leave, union officials said.

Employees were told of the bacteria the day the results were found, Egerman said.

The bacteria causes Legionnaires' disease—a deadly form of pneumonia that occurs in about 5 percent of people who come in contact with the bacteria. The bacteria also can cause Pontiac Fever, a flu-like ailment, according to Occupational Health and Safety Administration medical officer Michael Montopoli.

People become sick from the bacteria by breathing contaminated water in a mist form, such as in showers, humidifiers and sinks.

The bacteria grow in pools of water of between 70 to 120 degrees in temperature and usually develops in heating and ventilation systems, OSHA officials said.

The bacteria can be killed with chlorine or by temperatures greater than 160 degrees. Both measures were taken in the latest outbreak in the Richmond building, Egerman said.

"As far as we know, everything should be OK," he said.

A second test result will be available in early May, Egerman said.

The problem is that the water in the building's bathroom sinks is set at a tepid temperature, said Dave Mack, president of AFGE Local 1112, which represents 950 employees of the building. The water does not get hot or cold enough to keep the bacteria from growing, he said.

SSA and OSHA are working together to develop a long-term strategy to keep the bacteria out of the building, Mack said.

Options include regular chlorine and hot water treatments; plumbing the restrooms to mix hot and cold water at the sink; treating city water as it comes into the boiler or run only cold water into the sinks, he said.

Mack said the union hopes to go back to monthly, rather than quarterly, tests.

"This shows the protocol of regular testing was working," he said of the latest bacteria discovery.

OSHA has had "quite a number" of incidents of the bacteria in public buildings and addresses it in a proposal for Indoor Air Quality standards, spokeswoman Cheryl Brolin said.

This is the fourth time since the 1991 outbreak that the bacteria has been found in an SSA building, Egerman said. It has also been found in the Philadelphia and Chicago payment service centers and the Albuquerque, N.M., data operations center, he said.

All large SSA complexes are tested regularly for Legionella, an agency spokesman said.

Organisms

Larger than the broad class of microorganisms, larger animals are potentially encountered in water supplies. Larvae of insects and worms are found in surface water supplies but usually the mechanisms used to screen out turbidity screen out these larger organisms as well.

SAFE WATER

From all of this discussion, it can be seen that the topic of safe water or safe drinking water becomes a matter of subjective judgment among regulating officials. In general, drinking water supplies in the United States and Canada are safe for everybody.

Suffice to say that regulatory standards for safe drinking water exist. The standards specify safe levels of the chemicals and microorganisms in water supplies. Each standard has a published test method that has been agreed upon by a consortium of medical professionals and scientists as the acceptable method to determine the amounts of impurities.

LABORATORIES

The book *Standard Methods for The Examination of Water And Wastewater,* available from the American Water Works Association (see Chapter 19) is the main reference document used by Water Quality Laboratories throughout the United States and in some areas of Canada. These standards are published jointly by the American Public Health Association, The American Water Works Association and The Water Environment Federation. The volume is divided into 10 parts that address physical and aggregate properties of water; metals; inorganic non-metallic constituents; aggregate organic constituents; individual organic compounds; radioactivity; toxicity; microbiological examination; and biological examination.

Each part of the book details a standardized test method to determine if impurities exist in the water. A facility manager who desires to have the water tested will usually be quoted lab tests based on these "standard methods."

In general, a facility management professional concerned about the quality of the water supply under his jurisdiction will utilize a certified laboratory to test and verify the adequacy of the water. Bacteriologic monitoring costs between $200 and $300. Volatile organics tests can be expected to cost between $200 and $500. Synthetic organics tests are more expensive, around $2,500. Radionuclides tests cost under $200.

The facility manager's budget should include funds for testing on a periodic basis. Reporting should include the test results.

For each of the impurities discussed in this chapter, there exists a technology for removal. In general, a water system that is filtered and is treated with chlorine will meet the requirements for safe drinking water. For dissolved salts and other minerals, reverse osmosis or ultra filtration can remove the bulk of the impurities. For more discussion of the techniques and equipment used to purify water, refer to Chapter 11.

PUBLIC NOTIFICATION AND RIGHT TO KNOW

Finally, the facility manager should be aware of the legal requirements and the public's right to know. If the maximum amount of contaminants allowable drinking water standards are exceeded, then the facility manager must notify the public of the violations (see Chapter 3).

Chapter 5

Scale and Corrosion of Water Systems

*S*cale is one of the more common problems a facility manager encounters. Scale is commonly encountered during maintenance and the facility manager must determine if the scale is detrimental or not. Like scale, corrosion can be encountered during maintenance. This chapter will help the facility manager to determine if corrosion is significant and what the next steps are to solve the problem.

IMPURITIES

Both scale and corrosion are the results of water system impurities acting on the water system. In the case of scale, a deposit is formed inside the pipes, tanks or equipment. In corrosion, the impurities in the water are carrying away some of the system, usually metal. Scale can build up until it chokes off the pipe flow, or the solids can break loose and jam pumps or strainers. Corrosion can also lead to erosion, or to leaks in the system, which will cause a shutdown of the system while the leaks are fixed. To solve the problem find out what is causing the impurities and treat the water to remove or immobilize them. An alternative is to utilize a different piping material that inhibits the growth of scale or prohibits corrosion.

SCALE

Scale is a chalky white material found on the inside of pipes and tanks and is most commonly associated with heating equipment. Scale takes several forms but the most common is from calcium carbonate (CaCO3.) Calcium carbonate is present as an impurity in all water as the result of water slowly dissolving calcium from the environment over a long time. In general there is more calcium carbonate in well water than in surface water. See Figure 5-1.

Other impurities that form scale include calcium bicarbonate, calcium sulfate, magnesium chloride and magnesium bicarbonate. When these materials are present in the water and are deposited (or plate out) on the equipment, scale forms. They are relatively insoluble in water. All scale present in water before being deposited is called "hardness." Hard water has a lot of scale impurities while soft water has less. Generally, then, hard water will form more scale than soft water. Hardness in water is explained more fully in Chapter 4.

UNITS

The units to measure scale-forming impurities in water are referred to in milligrams per liter (mg/L). Since most water is of

Figure 5-1. Scale formation can affect piping flows when it builds up as it has in these two pipes.

density close to one, the milligrams-per-liter figures are comparable to parts per million or ppm. Hard water is usually defined by scale forming impurities in excess of 220 ppm. Soft water is usually below 100 ppm. Between these ranges the water is considered neither hard nor soft. There is an old term, still used, for hard water measurement called "grains" or "grains per gallon" and some hardness test kits give results in grains per gallon instead of parts per million. To convert hardness in grains per gallon to parts per million multiply the grains per gallon by 17.2 ppm/gpg.

Other salts in water that easily dissolve, such as sodium chloride, (common table salt) are not scale forming and will readily rinse away with more water. Scale, after it has formed must be removed by more stringent methods.

SYMPTOMS OF SCALE

Scale forms on the inside of water equipment, usually worse in heating and boiler water than in raw water or treated cold water systems. Scale buildup can be heavy enough to close off and affect the flow of pipes but more often the scale acts as an insulating barrier between heating equipment and the water. The scale acts as an insulator and creates a need for more energy to heat the same amount of water. A small amount of scale inside a water heater can reduce the efficiency of the heater, causing the facility to pay more to heat the water than is necessary.

The impurities that cause scale can also affect laundry operations since more soap is required to achieve the same level of cleanliness. In large laundry operations, water softeners can reduce the need for soap enough to pay for the water softening equipment.

Special applications such as laboratories or process/manufacturing plants may find that scale impurity may affect the process and interfere with the lab results or the process product.

Scale can break loose and plug pumps, pipes or strainers or the small pieces of scale can rub and erode the pipes or equipment and cause leaks.

THE LANGELIER SATURATION INDEX

Scientists and water professionals have found a way to determine if scaling will be a problem by determining if the water is saturated with calcium carbonate. If it is high, it will tend to plate out on equipment and piping, if low, then the water will tend to dissolve any scale formed. The Langelier Saturation Index generates a number that indicates scale or corrosion forming tendencies based upon impurities in the water. The ideal number is zero meaning it is not scale forming or corrosive. If the Index is greater than zero the Langelier Index indicates water has impurities that will form scale, if near zero, the water is stable and if less than zero the index indicates the water is corrosive.

How to calculate the
Langelier Saturation Index (LSI) for water.

STEP 1. Perform the following tests and obtain the results. These results are needed to calculate the Langelier Saturation Index or LSI.

Hardness in milligrams per liter

Alkalinity in milligrams per liter

pH of the water

Total Dissolved Solids in milligrams per liter

Temperature of the water

STEP 2. Calculate the saturation pH of the water using the formula below.

pH_s = A + B – log Ca+2 – log M

Where pHs is the Saturation pH of the water

A is an adjustment for the temperature of the water from the Table 5-1.

B is an adjustment for the Total Dissolved Solids from the Table 5-1.

log Ca+2 is the logarithm of the Hardness in milligrams per liter

log M is the logarithm of the Alkalinity in milligrams per liter

STEP 3. Subtract the Saturation pH of the water calculated in step 2 from the measured pH collected as data in step 1.

pH – pHs = Langelier Saturation Index (LSI)

If this number is greater than zero the water is scale forming. If it is less than zero it is corrosive. If it is at or near zero, it is neither scale forming or corrosive and is considered stable.

Example:

Step 1. Obtain the following Test Results

Hardness in milligrams per liter =	360 mg/L
Alkalinity in milligrams per liter =	60 mg/L
pH of the water =	8.3
Total Dissolved Solids =	650 mg/L
Temperature of the Water =	53°F

Step 2. Calculate the Saturation pH of the water using the formula

$pH_s = A + B - \log Ca^{+2} - \log M$

$pH_s = 2.30 + 9.87 - \log 360 - \log 60$

$pH_s = 2.30 + 9.87 - 2.56 - 1.78$

$pH_s = 7.83$

Step 3. Subtract the Saturation pH of the water calculated in Step 2 from the measured pH collected as data in Step 1.

$LSI = pH - pHs$

$LSI = 8.3 - 7.8$

$LSI = + 0.50$

In this example, since the Langelier Saturation Index is greater than zero, it would indicate this water would tend toward the forming of scale.

---

*Author's note: *http://www.awwa.org/Science/sun/langelier.cfm* has a Langelier Saturation Index Calculator that got the same answer in about 1/4 the time of preparing this calculator.

By using the Langelier Saturation Index the facility manager is able to determine if the water will be scale forming or not. The next step would be to neutralize or treat the water to prevent scale buildup.

Table 5.1 Values Necessary to Calculate the Langelier Saturation Index.

Temperature						Dissolved Solids			
°F	°C	A	°F	°C	A	mg/L	B	mg/L	B
32	0	2.60	104	40	1.71	0	9.70	500	9.86
41	5	2.47	113	45	1.63	25	9.73	600	9.87
50	10	2.34	122	50	1.55	50	9.76	700	9.88
59	15	2.21	131	55	1.48	75	9.78	800	9.89
68	20	2.10	140	60	1.40	100	9.80	900	9.90
77	25	1.98	149	65	1.34	125	9.81	1000	9.91
86	30	1.88	158	70	1.27	175	9.82	1100	9.92
95	35	1.79	167	75	1.21	225	9.83	1200	9.93
			176	80	1.17	300	9.84	1300	9.94
						400	9.85	1400	9.95

TREATMENT OF SCALE AND WATER WITH SCALE IMPURITY

Once the facility determines the water impurities have a scale problem the next step is to decide what to do about it. Reaming or scraping can remove hardened scale but this is labor-intensive work and the facility manager would want to decide if hand removal is the best method. If, for example, a decision is made to remove hot water scale from inside a boiler, it will be necessary to remove the scale from a confined space. Brushes, reaming tools, scrapers, and chippers will be needed and if the decision is made to remove the scale by hand, care must be taken not to damage the metal.

At many facilities a decision is made to replace the piping. This can be a costly decision, especially if the piping is located where access is difficult.

A third option is to chemically remove the scale using a mild acid like muratic or citric acid. Also, stronger acids can be used, like dilute nitric, hydrochloric, or sulfuric acid but care must be taken to preserve the metal. Too much acid can dissolve the metals. In the event of chemically dissolving the scale, a consultant or chemical treatment specialist is recommended. If a stronger acid is used, it will be necessary to treat the equipment with a basic solution after treatment to neutralize the acid. The facility manager must also be cautious of disposal of the chemicals used in the treatment since acids cannot be flushed down the drain.

Does magnetic water treatment really work?

Residential users and some small commercial users in areas with hard water and scale have been approached with the idea that a magnet on the incoming water supply line will reduce scale. Several attempts have been made to validate this claim and a test by Consumer's Reports in 1996 between two identical hot water heaters yielded no specific results. However, others who have installed the same equipment have reported that magnetic treatment does remove scale. One paper reported that the calcium carbonate crystals took a different form and did not stick and this reduced, but did not remove, the scale. No ill health effects have resulted from any tests with magnets installed to remove scale in hot water systems.

CORROSION

A more serious but less often problem encountered by the facility manager is corrosion of the piping or equipment in the plant. Corrosion is removal of material from the system and can lead to leaks that cause the system to be shut down while the leaks are repaired. Corrosion occurs when the water and pipe material (usually a metal) create a small electrical charge that removes molecules from the metal system. This corrosion can occur on the inside or outside of the system. Metal pipe that has been buried in the ground shows a common example of corrosion. Rust is a form of corrosion in steel/iron piping systems and components. Copper piping and equipment corrodes, creating small leaks and on occasion the complete failure of a joint or fitting. The result is a massive leak that soaks everything in the room.

Facility costs resulting from corrosion can be excessive. Utilities who have not understood the corrosive nature of their water systems have had to dig up miles of pipe systems. They have had to replace one metal pipe system with another, and they have had to deal with angry customers because of leaks in the system.

SYMPTOMS OF CORROSION

Corrosion is difficult to identify because it usually takes place over a long period of time and in a sense is an insidious process. Corrosion of steel piping can be seen in the form of rust colored stains where water is regularly drained. The water runs off or evaporates leaving the small amount of iron oxide or rust. Copper corrosion from copper sulfate formation leaves a blue or blue-green stain. Pinhole leaks in fittings or at valves is another indication of corrosion. Finally, bubbles or pits on the inside or outside of pipe are indicative of system corrosion.

TESTS FOR CORROSION

Corrosion results when the pH of the water is lower, when the Langelier Saturation Index is less than 1, or when impurities in water such as sulfur or carbon form mild acids that dissolve the pipe system.

Generally, water with the following properties influence the corrosion rate.

1. Water below 60 milligrams per liter hardness.

2. Water with a low pH. Less than 7.0

3. Water with high chloride or sulfate content. Greater than 150 milligrams per liter.

4. Water with large amounts of dissolved oxygen.

5. Water with high conductivity. Greater than 500 micro Siemens per centimeter.

6. Water with free chlorine greater than 1 milligram per liter.

7. Water with suspended solids such as sand or dirt.

The simplest non-destructive test for pipe corrosion is through ultrasonic testing. Ultrasonic test equipment can measure the pipe thickness from the outside using sound waves. The problem with ultrasonic testing is that it cannot accurately measure pitting. Heating equipment, especially steam boilers, are ultrasonically inspected.

Another method of testing and inspecting the piping system is through the use of a borescope or camera. This method is discussed in Chapter 17.

Another test for corrosion is through the use of test coupons that are inserted into a test apparatus and subjected to the water flow for a long period of time. These coupons are removed and measured to determine if any corrosion has taken place, usually by weighing them. The difference in weight reflects the amount of corrosion. This method of corrosion testing is slow and is better suited for pipe and material manufacturers, but if there is an indication that a corrosion problem can be suspected, this is the most effective test method. The time required to perform the test is several months. The test method and apparatus are defined in ASTM D2688 Test Method B.

However, another method for determining the amount of corrosion is inspecting the piping or equipment. One method would be by removal (and replacement) of a test section and having the removed section analyzed in a laboratory.

Steel Pipe

Two types of pipe, plain steel and galvanized steel are subject to pitting corrosion. Galvanized pipe is steel pipe that has been coated with zinc to protect it and prolong its service life. It has a light gray color. Plain steel pipe is often referred to as black iron pipe. Most pipe has its manufacturing standard stamped on the side and this can be read by facility maintenance personnel. In new facilities documents are usually turned over to the owner showing the materials of construction.

Pitting corrosion can be calculated based upon the known material of construction and water system impurities. Most pitting calculations give results in mils per year (mpy). A mil is one thousandth of an inch, so corrosion calculations are small. A method for estimating pipe pitting corrosion uses the Ryznar Index.

How to Calculate the Ryznar Index for Pitting Depth

The Ryznar Index is used to calculate pitting corrosion depth. Like the Langelier Saturation Index the Ryznar Index is used to calculate scale and corrosion tendencies in water and piping systems.

The Ryznar Index is a measure of the amount of calcium carbonate in saturation in water as opposed the actual amount. If the index is above 6, corrosion tendencies exist.

The Ryznar Index:

$RI = (2 \times pHs) - pH$

The Pitting Coefficient is determined from the Ryznar Index by:

$Pc = 0.0200 \times (RI - 7)$ for cold water

$Pc = 0.0261 \times (RI - 7)$ for hot water. Hot water less than 135 Degrees Fahrenheit.

Pit Depth is the Pitting Coefficient times the time in years to the 1/3 power.

$P = Pc \times t^{1/3}$

Where: pHs is the saturated pH of water. (The same as for the Langelier Saturation Index)

$pHs = A + B - \log Ca+2 - \log M$

pH is the measured pH of water.

RI is the calculated Ryznar Index

Pc is the Pitting Coefficient.

P is the pit depth in inches.

t is the time in years.

Also in the evaluating the corrosion of steel and galvanized steel, a soluble copper impurity in the water has been shown to increase pitting and corrosion in galvanized steel piping systems.

Copper Pipe

Like steel and galvanized steel pipe, copper pipe is also subject to corrosion. In recent years, copper has been selected over galvanized steel as it generally provides better corrosion and scale inhibition than steel. Copper has its own unique tendencies toward corrosion with free chlorine above 1 milligram per liter increasing the risk of corrosion. Also the presence of chloramine, another disinfectant, above 2 milligrams per liter can cause corrosion of copper pipe. Pitting of copper can also be a problem in water of pH below 7.8 and containing more than 17 milligrams per liter sulfate. Pitting can also occur in soft water of low pH (hardness less than 60 milligrams per liter and pH less than 6.5.)

Other symptoms leading to corrosion of copper are:

1. Poor soldering techniques and improper use of flux during the soldering process.
2. High temperatures of water: above 140 degrees Fahrenheit.
3. Manganese in excess of 0.05 milligrams per liter.
4. Sulfide in excess of 0.1 milligram per liter.
5. Presence of iron oxide or other insoluble matter
6. Ammonia in the water supply

A combination of high water temperatures and high velocity (excess of 140 degrees Fahrenheit and velocity exceeding 4 feet per second) lead to an erosion/corrosion problem in copper piping.

A method for computing corrosion of copper piping was developed by the Illinois State Water Survey using test coupons.

The corrosion rate (milligrams per square decimeter per day) = 2.993 − (0.03084 × milligrams per liter of carbon dioxide) + (.001857 × (mg/L Total Dissolved Solids) − (0.3268 × pH)

To convert this to mills per year multiply the result by 0.16.

Evidence of copper material corrosion is seen in a green color of water, unpleasant taste and green staining of plumbing and piping fixtures.

TREATMENT

Once the tendency toward corrosion is established, the facility manager has a number of options for treatment that can prevent further corrosion. Treatment of corrosion is a function of water impurities and the materials. Copper pipe, steel pipe and galvanized pipe are each subject to different methods of corrosion attack and the treatment has to be tailored to both the water quality and the material corrosion methods.

First, the facility manager can change the water chemistry by adding chemicals that reduce the corrosion. An example would be if the water has been softened to near zero mg/liter it may be possible to bypass a portion of the softener with hard water to bring the hardness back to a more agreeable parameter. The water can also be treated with corrosion inhibitors such as sodium silicate.

A more complex solution but one that does not impact water chemistry is to install cathodic protection of the piping system. This type of application is common for treatment of buried pipe. Cathodic protection is the installation of sacrificial metals designed to corrode and prevent the corrosion of the piping materials.

The metals can be coated or painted as a protective measure. Finally, the materials of construction can be changed to accommodate the corrosion problem. These later methods are more costly so a piping design that addresses corrosion initially is beneficial to the facility manager.

Treatment of copper corrosion can be accomplished by addition of sodium silicate to 4 to 8 milligrams per liter and raising the pH to 8.0. Allowing some calcium carbonate (hardness) to bypass the water softeners to maintain some protective qualities is also an effective method of corrosion treatment.

SUMMARY

Both scale and corrosion can be controlled by the facility when the water chemistry and amount of impurities are known. The Langelier Saturation Index and the Ryznar Index can be used to calculate scaling or pitting problems in Steel and Galvanized pipe. The Illinois State Water Survey has developed a method for determining approximate copper corrosion method. The main problem with scale is it will increase the facility energy cost. Corrosion can lead to piping and equipment failures that can be costly to the facility.

Now that the Facility Manager understands the corrosion and scale problem, the next step in Water Quality and Systems Management is to understand how to manage upgrades and renovations.

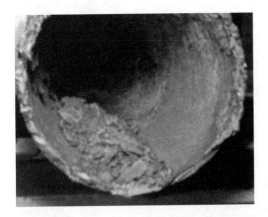

This 40-year-old sample of 8 in. schedule 80 pipe, while clearly containing deposits of iron oxide, shows very even wall loss and long remaining service life. The pipe was cleaned using high-pressure water jet and returned to service with approximately schedule 40 thickness remaining.

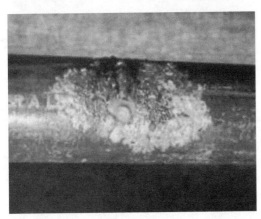

Pitting typically shows itself first at the smaller diameter piping simply due to the lower wall thickness present. Such evidence should be taken as an advance indication that a system wide problem may exist.

Figure 5-2. Corrosion can lead to piping leaks and equipment failures.

Chapter 6

Upgrades and Renovations

*M*any factors prompt an upgrade of the water system, such as changing user needs, controlling costs and replacing old, unreliable pipes and components. But other factors come into play as soon as the job is started—user demands, frustrations, a low budget and other factors—which may cause the water system to be upgraded without planning. As in any renovation or upgrade, careful planning is essential to success.

RENOVATION STRATEGY

In facility water management, the time comes when the manager decides it is time to upgrade the system. The decision should be a logical one, driven by the life of the system and the scheduled replacement time of the system's components.

While a simple residential bathroom remodeling project is relatively inexpensive, as low as $3000 in 2003 dollars, remodeling the bathing facilities for a large airport or sports arena is a major task, involving new subgrade, pipes, plumbing, fixtures, flooring, wall tiles, lighting and air conditioning.

Toilet facilities, because of the coordination of trades and the cost of the materials, are one of the most expensive elements. In addition, because they are not rented, their cost is incorporated into the cost of the lease. Hence a facility manager who manages a commercial rental property finds that one of his most expensive rooms in the building generates him the same revenue per square foot as general spaces.

Conduct A Needs Assessment

The first step in the upgrade process is a needs assessment, Essential questions to be included in making an assessment as to whether to upgrade or not are listed below. In answering the questions, if the response is not known, provide a best estimate. It is possible to go back and fill in the details a little later.

1. How many maintenance work hours have been spent repairing or replacing plumbing and piping in the facility in the past year?

2. What is the total dollar value of the materials that have been installed in the past year?

3. Have there been any written complaints about the water supply? Review records of complaints.

4. How old are the plumbing fixtures in the restrooms?

5. Has the facility been contacted by the city and notified of the requirements to notify the public about health hazards in the existing water supply?

6. Any flooding or wastewater problems? What were they?

7. Has the facility staff made suggestions for improvement to the facility that would improve service?

8. Have new standards or regulations identified any potential requirements to change or modify the water supply system?

In addition, the facility manager should consider changes to the use of the space since the original water system's installation, and whether the restroom fixtures, tile and other features are pleasant and compatible with the current decor or image of the general spaces. This can be particularly critical in high-profile business such as hotels and restaurants, but corporations may consider appearance critical as well for visitors and employees alike. Additionally, if the upgrade incorporates efficient electric motors, controls, low flow devices or water-saving methods, we must also consider potential energy and water bill savings in the selection of components and design. All of these benefits must be weighed against cost.

For a hotel, for example, the basic question would be: If we remodel these rooms, how much will we be able to increase the rental rate? If we cannot increase the rental rate, will we be able to increase occupancy? Can this formula be tested? How? It is easy to estimate costs. It is quite another problem to estimate yields, especially one that is accurate.

Conduct A Utility Survey

The next step is to conduct a survey of the water system. We want to know what types of equipment are in operation, find out how much the water system costs in energy and water, and generate a profile of water use and balance between supply and wastewater.

Is the water coming into the facility being matched by the water going out? As discussed in Chapter 1, we can construct a model of water use based on the monthly utility bills; these bills for supply water can be used as an initial guess of the water going out.

The manager should validate if the water meters are accurately indicating the flows. If the utility readings are based on the flow meter, this is adequate for an initial assessment. Later, the facility may decide to have the flow meters calibrated to verify the accuracy of the flows.

If older bills are available in the records, a pattern can be established offering seasonal use and flows. New York City once estimated that approximately 21 percent of its water supplies is unaccounted for in utility bills.

Next, water use can be estimated based upon the types of items and number of people using water at the facility. Figure 6-1 is a form that allows the facility to collect the necessary data for finding out where the water is being used. It is desirable, in order to keep track of which worksheets go with which facilities and spaces or rooms, to use the worksheets with a floor plan, numbering each facility and room.

Figure 6-1. Form for performing a water utility study.

Water Survey   Sheet _____

Name of Facility _____   Location of Facility _____

Name of Person Collecting Data _____   Phone _____

Work address   _____

Room Number or Name   _____

Number of Water Closets _____ Manufacturer/Brand name _____

Number of Urinals _____ Manufacturer/Brand name _____

Number of Showers _____ Manufacturer/Brand name _____

Number of Sinks _____ Manufacturer/Brand name _____

Are there water entertainment/recreation items at the facility?

      Swimming Pool_____ Yes   Size _____ Gallons _____ Sq.Ft.

      _____ No

      Fountain   _____ Yes   Size _____ Gallons _____ Sq.Ft.

      _____ No

      Jacuzzi/Spa   _____ Yes   Size _____ Gallons _____ Sq.Ft.

      _____ No

Hot water boilers   _____ Yes _____ No

Steam water boilers _____ Yes _____ No

Natural gas boilers _____ Yes _____ No

What kind of fuel is used for heat?

Natural gas   Cost: _____

Fuel oil   Cost: _____

LPG gas   Cost: _____

How much does water cost for the facility? $ _____ per _____

Are there water treatment systems such as water softeners? _____ Yes _____ No

    Make _____   Model _____

    Water used per cycle _____ Gallons* No. cycles per day _____

Is there are a laundry?   _____ Yes _____ No

    Number of washing machines _____

    Make _____   Model _____

    Water used per cycle _____ Gallons*   No. cycles per day _____

Are there any other items consuming water?   _____ Yes _____ No

    What? _____

    Make _____   Model _____

    Water used per cycle _____ Gallons* No. cycles per day _____

Hint: Check the operations and maintenance manual if you have one; if not, get the make and model and call a local vendor and ask him.

Constructing a Model of Water Use

Once a tally of all the devices has been completed, Table 6-1 can be used to fill out the number of gallons of use per fixture per day. By extending the numbers of fixtures times the flows per fixture a total of flows for that type of device can be calculated per day. This can be extended into a total of all flows per day.

Now that an exact count of the devices and an estimate of the flows is known it is possible to check the calculated totals from fixture use against utility bills and to note any significant differences. Managers who are familiar with spreadsheet formulas on personal computers will find this type of information lends itself quite readily to a spreadsheet analysis.

Table 6-1. Minimum flow and pressure required by typical water-using devices, used for estimating flow and constructing a model of facility water use. Source: U.S. EPA, *Manual of Individual Water Supply Systems*, 1975.

	Flow Pressure*		Flow Rate	
	(psi)	(kPa)	(gpm)	(L/s)
Ordinary basin faucet	8	55	2.0	0.13
Self-closing basin faucet	8	55	2.5	0.16
Sink faucet, 3/8-in. (9.5 mm)	8	55	4.5	0.28
Sink faucet, 1/2-in. (12.7 mm)	8	55	4.5	0.28
Bathtub faucet	8	55	6.0	0.38
Laundry tub faucet,				
1/2-in. (12.7 mm)	8	55	5.0	0.32
Shower	8	55	5.0	0.32
Ball-cock for closet	8	55	3.0	0.19
Flush valve for closet	15	103	15-40**	0.95-2.52**
Flushometer valve for urinal	15	103	15.0	0.95
Garden hose (50 ft., 3/4-in.				
sill cock) (15 m, 19 mm)	30	207	5.0	0.32
Garden hose (50 ft., 5/8-in.				
Outlet) (15 m, 16 mm)	15	103	3.33	0.21
Drinking fountains	15	103	0.75	0.05
Fire hose 1.5-in. (38 mm)				
1/2-in nozzle (12.7 mm)	30	207	40.0	2.52

*Flow pressure is the pressure in the supply near the faucet or water outlet while faucet or water outlet is wide open and flowing.

**Wide range due to variation in design and type of closet flush valves.

The number of fixtures and total flows calculated from extending the fixtures times the flows is called a "water use model."

Adjusting The Model to Meet The Flows

Once built, the model should be adjusted for actual versus calculated totals. This process is called reconciliation. Using the known information from the utility bills, the estimated flows from the model are compared to the actual utility meter readings. This allows the manager to verify the accuracy of the model. If flows calculated are within 15 percent of the amount indicated on the utility bills, the results of the model can be considered fairly accurate.

If the differences a greater than 15 percent, the model can be "ratioed." (see Figure 6-2). By ratioing, the model can be made to correlate with the utility bills.

Figure 6-2. Ratioing the water use model—a technique used to reconcile known metered water use based on utility bills with the facility's water use model if a difference in flows greater than 15 percent exists.

Actual Use (from Water Bills) ÷ Calculated Use (from Model) Ratio*

Calculated Use		*Ratioed Use*
Ratio x Sinks	=	Sinks Use
Ratio x Building A	=	Building A Use
Ratio x Lawns	=	Lawns Use
Calculated Use Total	=	Actual Use
Example: Water Billed	=	28,265 gallons (Actual Use)
Calculated Use	=	25,200 gallons
28,265 ÷ 25,200	=	1.12
1. 12 x Sink Use	=	Sinks Use
1. 12 x Building A Use	=	Building A Use
1. 12 x Lawns Use	=	Lawns Use
Total	=	Actual Use (28,265 gallons)

*The ratio should not be more than 1.25. If it is, the model has likely missed a major user not yet identified. Check the facility to make sure that all water use has been accounted for. Some places where water use is missed is in mechanical rooms or boiler rooms where water is used for cooling. In addition, a leak could be present in the lines at some point.

The adjustment in the ratio is usually a result of a facility pressure different from the values assumed in Table 6-1. If, in Table 6-1, the flow was estimated at 20 gallons of water per day and the pressure at the facility is less than 60 pounds per square inch (psi), then perhaps the flow at that facility is only 15 gallons per day (gpd). If, on the other hand, the pressure is 120 psi, then the flow is going to be 30 gpd. So ratioing is not an uncommon method for adjusting models.

Next, the facility manager should locate a map of the utility piping. This map will show where the water lines are and it may be possible to isolate portions of the facility and measure flows to that area. If each building is submetered in addition to a meter for the entire complex, the individual meters can be used to total the flows for the entire complex. Do the flows total the same?

Remember water is not terribly expensive, so a little difference between the totals is not going to matter. Small leaks, drips or the gardener using a hose to wash down the sidewalk should not be enough to cause concern.

Another way to measure flows is to allow the device to flow into a known volume and measure the time it takes to fill the volume. For example, if it takes two minutes to fill a 5-gallon pail, then the flow can be estimated at 5 gallons ÷ 2 minutes = 2.5 gallons in one minute, or 2.5 gpm.

Analyze the Model

Once the flows are complete, the model's results can be analyzed to see where the flows are going.

Where is most of the use? Is it in the restrooms? The water heating? Is water being needlessly wasted?

By studying the water use model, the facility can determine how to save both water and money. In addition, hot water and boilers for utility processes and for facility heating can be analyzed and estimates for energy costs established as well.

Compare Alternatives and Decide

Just by knowing where the water is going is effective at showing the facility manager what his present water costs are. Once the facility has a model and the manager is confident that it is representative of actual water use in the facility, economic comparisons of alternative water uses are made.

The U.S. EPA's WAVE Saver Program

The U.S. EPA has formed Water Alliances to Save Energy (WAVE) to help industry economize on water and energy use. A membership in WAVE entitles partners, supporters and endorsers to utilize the U.S. EPA's promotional material and the WAVEsaver. WAVEsaver is a computer modeling program like the ones discussed in this chapter.

By joining WAVE, facilities in the hotel/motel industry can request the program and use it for modeling and tuning water use.

WAVEsaver program materials include an instruction manual, guideline forms and a CD-ROM to install the system on an IBM-compatible computer. The program requires about 8 megabytes of RAM and 15 megabytes of space available on the hard drive.

The WAVEsaver program also comes with the necessary forms for conducting the utility field survey which is entered into the computer program before analysis is run.

WAVEsaver calculates the true incremental cost of water, creates budget projections based on historical utility rates and occupancy patterns. The program contains hundreds of databases that enable the facility to conduct cost-efficiency options and to select and customize property-specific studies. Forms are available from U.S. EPA.

The CD-ROM version provides interactive video and sound to analyze high-efficiency system upgrades.

The WAVEsaver program was developed in conjunction with water use experts and jointly sponsored by the U.S. EPA and the Metropolitan Water District of Southern California.

For more information, call the U.S. EPA at (202) 501-2396 or on the internet: www.epa.gov.

Use Low-Flow Fixtures and Components

Water closets, urinals and water faucets can be improved to use less water.

Water conservation can be accomplished when upgrading by installing low-flow shower heads, aerators on sinks and lavatories, low-flow or ultra-low-flow toilets, low-flow urinals and other water-saving equipment.

Heating systems can be changed to more direct methods that reduce energy costs. For example, a direct natural-gas-fired water heater can be cheaper to operate than a steam water heater because of the energy lost in the steam system between the steam boiler and the hot water heater.

By taking the number of fixtures in the model and applying use factors from water conservation measures, a new scenario of water use can be generated. This can then be totaled against the utility bills to show the dollar value of the savings.

A good model will reveal where water can be saved, how much can be saved and total the value of the savings. In addition, the facility can capitalize upon the marketing value of being a good water conservation steward which helps with marketing the facilities' public image.

The savings from use of low-flow water fixtures and from implementing water savings measures can be used to off-set the costs of remodeling and may even repay the total costs of the improvements.

Chapter 7
Pipe and Fittings

*P*ipes and the associated equipment form the heart of any water system. It is pipe and pipe fittings that transport the water to where it is used and carry it away again when it is wasted. Knowledge of water systems and their management cannot be successfully accomplished without a basic understanding of pipe, piping and pipeflow.

PIPE TALK

"Pipe consists of a long hole, mostly tubular, usually straight, sometimes twisty."

So began a humorous specification once passed around by plumbers for a good laugh. In fact, pipe is one of the truly great inventions of our modern age. Water and wastewater systems depend upon pipe because the delivery and removal of water is the result of installing pipes to carry the flows.

Pipes come in lengths and are fitted together at joints. Depending upon the pipe and where it is placed, different types of pipe and joints are used. Joints, like pipe, are selected for ease of construction, maintenance, life and cost.

PIPING

Several excellent books exist for those who want to get really familiar with piping. Appendix 1 lists some of the books on the market. Since the facility is going to purchase pipe from a supplier, a discussion of how pipe is made is not presented in this text. Facility managers should know and understand that pipe comes in common sizes. When pipe size is discussed, the diameter of the end of the pipe is the number that is presented.

For supply, pipe line sizes vary from one inch for a single family dwelling up to large utility lines of 36 inches. Larger sizes, which apply to aqueducts and utility services, continue up to 120 inches (10 ft. in diameter). The smallest pipe is usually considered to be 1/2-inch. Below this size, pipe is usually referred to as tubing. For water management, the smallest size is normally going to be 3/8-inch and serves bathroom sinks and water closets.

Pipe sizes increase in uniform dimensions because the mills where pipe is made uses common size dies and molds. Commercial pipe comes in the following sizes: 1/2," 3/4," 1," 1-1/2," 2," 2-1/2," 3," 4," 6," 8," 10," 12," 14," 16," 18," 20," 24," 30," 36" and up in six-inch increments beyond 36."

For a long specialty-sized pipeline, a mill might be willing to set up a special run—for example, 20 miles of one unique size could be fabricated. However, most designers have found that specifying the next larger size is more economical. Consequently, pipes are sized according to these common diameters.

PIPE FITTINGS

For changing the direction of pipe and for hooking pipe up to fixtures, "pipe fittings" are used. Fittings are common and are standardized throughout the industry. Pictures of typical pipe fittings are shown in Figure 7-1. The three most common types of fittings are tees, ells, and wyes (T, L and Y). Just as the name implies, a "tee" is shaped like the letter T. Tees allow a pipes to be combined at a junction.

Like pipe tees, "ells" are used for pipe bends. All pipe ells are shaped like the letter L and allow direction changes in pipe. Ells can be set in the horizontal position, (looking aside), in the vertical position (looking up or down) or at other angles (halfway between up and aside). Ells for many systems can be either short radius or long radius. The advantage of long radius ells is there is less pressure lost through a long radius ell than a short radius ell. Usually drain systems use long radius ells, if there is room, to minimize the chance of plugging.

The Facility Manager's 10 Questions

There are 10 questions the facility manager should ask the design engineer:

1. How much excess capacity is there in the pipelines for later growth?

2. What piping materials have been selected, what are the advantages and disadvantages of each, and how long will the selected pipe materials last?

3. What pressure and flow tests have been specified and how will I know if the pipes pass the tests?

4. Has the engineer prepared a list of recommended bidders, or what experience is the engineer requiring of the construction companies?

5. Will the pipes be sterilized after construction? If so, where will the chemically treated water from sterilizing be disposed?

6. If the pipes need to be tapped in the future for new service locations, how hard will this be and what will be the procedure?

7. What special corrosion protection will be necessary? How is it applied?

8. Is the pipe material guaranteed? Who is furnishing the guarantee?

9. Where are the flow meters going to be? How hard are they to get to?

10. What happens if the meter breaks? Will the water have to be shut off while the meter is fixed?

In a similar manner to tees and ells, "wyes" are used to change the direction of pipes. Wyes, so named because their shape is similar to the letter Y, are also used to change the direction and in joining of pipes. Wyes are most often used in drainage pipes instead of tees, because the direction change is more gradual. In addition, wyes allow drain unplugging tools to follow the direction of flow.

One rule about fittings is that they must be the same material as the pipe—otherwise, temperature and corrosion will cause a failure of the pipeline between the two dissimilar materials. For a change in materials, a special fitting sometimes called a dielectric union is used. These specially fabricated fittings are designed especially for transitioning from one type of pipe material to another.

A "coupling" is used to join two lengths of straight pipe together. On some types of pipe systems, couplings are necessary. On others, the length or joint comes with one end ready to receive the end of the next one. The joints shown in Figure 7-1 are typical bell and spigot joints.

Finally, there are a number other specialty types of fittings. Plugs and caps close up the end of a pipe. (A plug is used to close a female end and a cap closes a male end.) Another common fitting is a reducer which is used to transition in size from one type to another.

Fancy fittings combine these major elements. For example, a tee can be a reducing tee where the sizes of the openings are different. Similarly, a reducing ell can change size from one end to the other. These specialty fittings are used for tight installations were there is not room to install two fittings. For example, a reducing ell would be used where there was not enough room to install an ell and a reducer.

Fittings can be purchased with any of the joints discussed later in this chapter, and many specialty fittings serve to change the type of joint from one type to another. For example, a fitting could be a bell on one end and screwed on the other. This would allow the pipe to be attached at one end to an appliance while the other has joints for the pipe. A typical example of this is where the water line taps into the common residential water heater. The water heater is usually a screwed fitting, while the copper pipe is

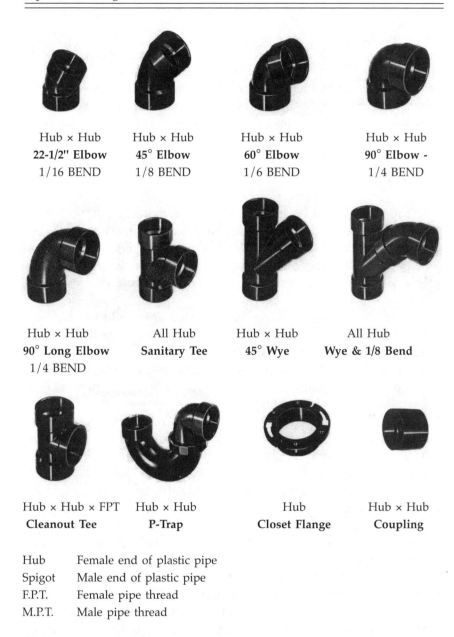

Hub × Hub
22-1/2" Elbow
1/16 BEND

Hub × Hub
45° Elbow
1/8 BEND

Hub × Hub
60° Elbow
1/6 BEND

Hub × Hub
90° Elbow -
1/4 BEND

Hub × Hub
90° Long Elbow
1/4 BEND

All Hub
Sanitary Tee

Hub × Hub
45° Wye

All Hub
Wye & 1/8 Bend

Hub × Hub × FPT
Cleanout Tee

Hub × Hub
P-Trap

Hub
Closet Flange

Hub × Hub
Coupling

Hub   Female end of plastic pipe
Spigot   Male end of plastic pipe
F.P.T.   Female pipe thread
M.P.T.   Male pipe thread

Figure 7-1. Pipe fittings: ells, tees and wyes. Reprinted from *Step by Step Guide Book on Home Plumbing* with permission of Step By Step Guide Book Co., West Valley City, Utah.

a slip fitting.

It can be difficult for the facility manager's staff to locate special fittings that will have multiple combinations such as a reducing ell that has a slip joint on one end and is screwed on the other. These types of fittings can be obtained but they are rare and sometimes may take a few days to locate.

PIPING MATERIALS

Pipe is made from every conceivable material and it is used to carry materials much more sophisticated than water. Pipes are designed and used to carry acids and even poisons. But for most facility managers responsible for water management, pipes carry water. For water, pipe types boil down to the few essential ones listed here.

Pressure pipes carrying fresh water are made of the least expensive materials that will contain the pressures and the flows. Water and sewer pipes are made from concrete, ductile iron, steel, cast iron, clay, white plastic (polyvinyl chloride/PVC), black plastic, (acrylonitrile butadiene styrene/ABS), blue plastic, copper and lead (lead being no longer allowed in the United States). Pipe choices should be made by engineering professionals familiar with this type of work, but the facility manager can direct the engineer to analyze the options and present them during initial studies if a renovation is being planned.

Overall, the system should provide the facility manager the optimum combination of price and function that results from judgment, analysis and experience of pipe life, cost, joint design, corrosion resistance, friction and pressure loss, hangers and supports.

For buried pipe, supply pipes are usually ductile iron. This pipe is strong, it resists water hammer, it is corrosion resistant, and it is flexible enough to bend slightly without breaking. Inside buildings, pressure pipes are made from steel or copper. The size at which copper becomes more attractive than steel is near the 3-4 inch diameter pipe size. This is because the smaller copper fittings are less costly and easier to install than steel fittings.

Case Study: Polybutylene Piping System Loses Lawsuit

In an article in the *Los Angeles Times* dated October 25, 1994, the manufacturers of gray supply piping agreed to pay damages stemming from a class action lawsuit filed in Texas. In the settlement, manufacturers of polybutylene pipe agreed to pay for replacement and damages resulting from leaks from polybutylene pipes used in homes, apartments and businesses nationwide.

According to the *Times* article, Shell Oil Company, DuPont Company, and Hoechst Celonese agreed to pay damages to members of the class who had used gray plastic polybutylene pipe advertised as a new product in the 1980s.

The problem with the pipe was that it was subject to chlorine attack and corroded, causing the leaks.

The settlement agreed to cover future leaks for up to 16 years after installation under certain circumstances.

If the facility has polybutylene plumbing or piping and it was constructed before 1995, leaks or damages from leaks may be reimbursable. Contact the Consumer Plumbing Recovery Center, Plano, Texas. 1-800-392-7591.

Ductile Iron Pipe

Ductile iron pipe is specially manufactured for use in underground water supply lines but it can also be used for sewer pipe. It offers a good combination of corrosion resistance, strength and price. Ductile iron pipe is relatively heavy compared to some of the other pipe types presented here, which is why it is not generally used inside buildings. Most ductile iron pipe is of the bell and spigot type, although special joints are sometimes used.

Steel Pipe

Steel pipe is also often used, both inside and outside buildings. Steel pipe in smaller sizes is usually screwed, or welded when in larger sizes. In general, steel pipe has excellent strength, but is more subject to corrosion than ductile iron pipe. Unless kept completely full of water, the inside of steel pipe rusts, resulting in the red color of water when pipes are initially turned on for a few moments. New steel pipe can be coated both inside and outside. Often, steel pipe can be identified on the job site because the outer coating is wrapped with paper. The most common steel pipe is referred to in the field as "black iron" pipe. Most black iron pipe is more specifically referred to as A53 Grade B pipe. This vernacular stands for pipe fabricated to the specifications in ASTM Standard A53, using the chapters for Grade B.

One thing to note about steels—they come in many alloyed forms. The facility manager should recognize that there are hundreds of other pipe and tube standards for steels. These other standards are for more special applications such as chemical and nuclear power plants. The specialty applications require more strength, corrosion resistance, thermal expansion and numerous other characteristics.

Common black iron pipe strengths are referred to by schedule. The most common is Schedule 40, but Schedules 20, 80 and 120 are also used. The difference in schedules refers to the thickness of the pipe wall, with Schedule 20 the least thick, next Schedule 40 and so on. It is quite common to see Schedule 40 black iron pipe used for water service and Schedule 80 black iron used for steam.

Copper Pipe

Like steel pipe, copper pipe is most commonly used in buildings for hot and cold water supply piping. Copper, while not as strong as steel pipes, is much lighter and easier to fit together than heavy steel pipe. In addition, the lighter pipe allows the use of lighter pipe hangers and supports which further reduce costs in this system. Copper pipe joints and fittings are usually brazed together using a process called "sweating." Because the copper pipe is thin-walled, its size is limited to about 4 inches in diameter. In addition, it is difficult to braze copper pipe in larger diameters. Copper pipe is rarely screwed, except where special fittings are used.

Copper pipe comes in special classes designated by letter, with the most common being type K, type L or type M. The thickest and strongest of these is type K. The outside diameter of each type of copper pipe stays the same so that the same fittings can be used for each type.

Finally, copper pipe does not work well in underground service because of the conductivity of copper metal.

ABS Pipe

Acrylonitrile Butadiene Styrene (ABS) pipe is a black plastic pipe used for drain and vent piping in buildings. ABS is used because it is lightweight and easy to cut, install and hang. Generally, ABS is not used below ground. It is also not used for outside service because sunlight's ultraviolet waves tend to break down the plastic and make it brittle over time. ABS pipe is used as a vent pipe to project through the roofs of many buildings. Since this vent pipe carries air and sewer gases, the brittlement from sunlight exposure is ignored for this short few feet of pipe. ABS pipe can be painted if necessary to protect it.

Since the pressures of drain pipes is nearly zero, compared to 20-90 pounds per square inch for pressure pipe, ABS is much cheaper than copper or steel pipe would be in the same diameters.

PVC Pipe

Polyvinyl Chloride (PVC) pipe is sometimes used for water supplies and is often the pipe of choice for lawn sprinkler systems. It can also be used for inside building supply lines. Joints and fittings of PVC pipe are glued together, although some special fittings are used to make screwed joints. As with copper, steel and ductile iron, PVC pipe comes in different strengths called pressure classes. The thinnest is called class 200 but the most common type used is Schedule 40. Schedule 20, Schedule 80 and Schedule 120 Plastic pipe is also available.

Schedule 40 PVC pipe is almost as expensive as steel pipe, but the advantage of PVC pipe is that it is easier to cut and glue together in the field than steel pipe, which has to either be welded or threaded and screwed together. In general, the fittings for PVC pipe are much less costly than fittings for steel pipe until the size gets up to about 6 inches in diameter. Then the steel fittings are less than the plastic fittings.

Other Plastic Pipes

There are several other types of plastic pipes that have advantages and trade-offs similar to the ones mentioned for steel, copper, ABS and PVC. These other more sophisticated pipes are generally identified by their color since manufacturers want to make it easy to identify their product in the field. The two most common to be encountered by a facility manager are made from blue plastic and orange plastic. Similar to PVC, blue plastic pipe is used for underground water supplies. The resins in this "blue brute" pipe have been designed to be more corrosion-resistive than the white PVC pipe. Orange plastic pipe is used as an alternative to Schedule 40 black iron pipe for fire sprinklers in residential construction. The advantage of the orange plastic has been that it is specially fabricated to be fire-resistive.

Fiberglass Pipe

Fiberglass pipe is often used as an alternative to steel pipe but in general it is more expensive. Fiberglass pipe is used more often for chemical and refinery operations. One example where fiberglass pipe might be encountered by a facility manager is in the acid piping used for regeneration of mixed bed deionizers (see Chapter 11).

Concrete Pipe

Concrete pipe is often used for underground sewer service, although it has been losing ground in recent years to PVC pipe. The big advantage to concrete pipe has been its resistance to buckling under roadways. The smallest size for concrete pipe would be 4 inches, with 6 inches being more common. Concrete pipe is also used for pressure pipe lines and there is a lot of competition between the concrete pipe manufacturers and the ductile iron pipe manufacturers to bid and install their types of pipe. A facility manager may be able to take advantage of this competition by instructing the engineering staff to prepare contracts in such a way that either type of pipe can be installed. One note of caution, however, has been the preference of utilities located in areas subject to earthquakes to prefer steel pipes over concrete for water supplies because of its flexibility.

Clay Pipe

Clay pipe is sometimes used for drainage applications instead of concrete pipe. Clay pipe can be less costly than concrete, particularly in areas where there are lots of clay soils and not much in the way of sands and gravel for making concrete. Clay is smoother on the inside than concrete pipes. Clay is used in some areas of the country, while in other areas it is rarely used.

There are many other pipe types that will commonly be encountered by water managers of facilities, and the facility manager should keep an open mind in considering options provided he can meet the city or state building codes.

Pipe in Older Facilities and Buildings

In many buildings, particularly buildings constructed using copper pipe during the 1940s-1960s, lead was used in joints or even for the pipe itself. U.S. EPA studies revealed that lead in small amounts can have adverse health effects on humans (see Chapter 4, concerning water purity). As a result, lead has been banned from piping materials. Water quality tests from the taps in the building or on the property will reveal if trace amounts of lead are present. (Water quality testing is also included with Chapter 4). Other old pipe materials include galvanized steel pipe, plain steel, and copper.

REMODELING/NEW PIPE

Many, many professionals in the engineering business make their living making pipe and there is not any way, without years of study, to learn this information quickly. The American Society for Testing and Materials, the American Society of Mechanical Engineers, and the American Water Works Association are a few agencies that write general material specifications for most types of pipes.

These standard specifications, some of which are several hundred pages long, define the materials, the stretch and break tests, the pressure tests, the penetration resistance, corrosion resistance and so on. The facility manager should make sure that the engineer designing the pipe system is familiar with these requirements when starting a major job.

WATER SUPPLY PIPING

Pipes types and fittings from the previous discussion are used to provide water supplies to buildings. Supply piping is treated and pressurized for delivery. Water supply piping keeps the water at the required pressures without leaking. The pipe is inert and does not impart any taste or smell into the water.

WASTEWATER PIPING

The previous discussion for supply pipe can be used to apply to drainage pipe as well. The two types of drain pipes in any facility are the stormwater drains and the sanitary sewage drains. The major difference between supply water piping and wastewater piping is that supply water piping is designed to flow full while wastewater piping, because it carries debris and other solids, is designed to flow only partially full (see Figure 7-2).

STORMWATER PIPE

Stormwater drainage runs down the curb and gutter, off the roofs of major buildings, down streets during summer rainstorms. It carries snowmelt and sometimes salt that has been added to roads, excess irrigation water from watering lawns and gardens, and in general simply allows the water that falls on the property in the form of rain to run off.

Stormwater designs are unique because they are really only used during those times of the year when rainfall or snowmelt occur. Some stormwater is channeled to ponds where it collects and evaporates, while other stormwater is channeled into rivers or gullies. Stormwater runoff carries its own types of problems which include debris and chemicals.

Stormwater will carry debris that can be moved by the flowing water. Wood, trash, paper and small stones are all types of

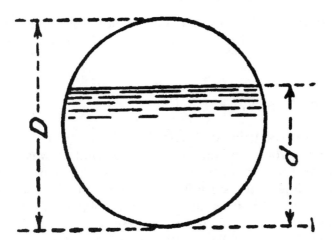

Figure 7-2. **Pipe partial-flow cross section. Courtesy: U.S. Bureau motion, Denver, CO,** *Earth Manual: A Guide to The Use of Soils as Foundations and as Construction Materials for Hydraulic Structures,* **First Edition.**

wastes carried by stormwater flows. These wastes, if carried into a neighboring property, have the potential to impact the facility manager and his operation. By concentrating the flows from a large complex into a single source, if water flowed onto an adjacent property as sheet flow it could have little impact on the neighbor. Concentrating the flow from a large area into a single pipe, however, can cause significant erosion over time that might require mitigation by the facility manager.

The second potential problem with stormwater piping is the chemicals that become mixed into the water as it runs off of the land. For example, if a lawn is chemically treated for preventing weeds with a commercial herbicide and rain causes the chemical to rinse off onto an adjacent property that is growing a crop, the chemical may kill or otherwise reduce the yield of the adjacent property owner's crop.

As a result of these problems, many facilities have special ponds designed to capture stormwater where it can the tested before being released. Other facilities have written understandings and agreements with the property owners downstream.

The same types of pipe used for supply water can be used for stormwater runoff but, because they are not designed for full flow, stormwater drain pipes will be larger in diameter than supply pipes.

SANITARY SEWER PIPE

Wastewater from kitchens, bathrooms and toilets carry washwater, garbage and human or animal sanitary waste. Because these sewer lines are for sanitary purposes, they are called sanitary sewer lines. Like stormwater piping, sanitary lines are also designed for partial flow and can carry debris such as garbage, excrement, sanitary napkins and paper.

The pipe materials are the same as the ones used for stormwater and since both stormwater and sanitary sewer flow partially full their sizes are larger than the supply water pipe for the same buildings or area served.

Along with the drain pipes flowing partially full, designs for drainage pipes keep them purposefully at zero pressure. In this way, a leak in the pipe will not flow out of the pipe as fast as it would if it were a pressurized pipe.

BURIED PIPE

Pipe buried below ground is usually placed with the drain pipe below the level of the supply pipe so that a leaking wastewater pipe will not contaminate the supply pipe. Having the supply pipe pressurized also prevents wastewater from getting into supply water. Occasionally, however, the supply water pipeline has to be drained in order to work on it. When a drained supply water pipe is filled, codes require the supply pipe to be sterilized (see Chapter 16).

DOUBLE-WALLED PIPE

Occasionally, although for water systems the need is rare, the facility will install double-walled pipe (see Figure 7-3). Double-walled pipe is more often associated with hazardous materials, but nonindustrial facility managers may benefit from knowing it is available. Double-walled pipe is essentially two pipe systems, one inside the other. The inner pipe is called the carrier pipe, while the outer pipe is called the containment pipe.

Double-walled pipe is expensive and is made from some elaborate materials like fiberglass and plastic. Steel is also used. A clip is used to keep the inner pipe separated from the outer pipe (see Figure 7-3). The area between the two pipes is called the interstitial space. Usually the interstitial space is air, and often an electronic cable is run through the interstitial space to determine and pinpoint leaks in the carrier pipe.

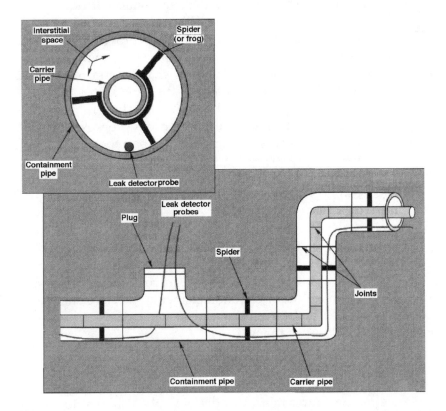

Figure 7-3. Double-walled pipe. Courtesy: *Heating, Piping and Air Conditioning Magazine,* **April 1993, original manuscript by the author.**

Corrosion

Pipes, especially pipes installed in the ground, are subject to corrosion from both the outside and the inside. Usually corrosion from the inside is minimal because water or sanitary waste is not highly corrosive.

From the outside, pipe is subject to corrosion from acidic chemicals in the surrounding soil. Pipe buried in the ground is subject to corrosion from a process called galvanic action.

Galvanic action is the result of electronic eforces on the pipe that cause the joints to become either positively or negatively charged. What happens, in effect, is that the pipe joint acts like a low-grade battery and over a long time the ends corrode in the same way that flashlight and car battery terminals corrode. Eventually, pinhole leaks develop in the pipe.

Corrosion in a pipe system is reduced by providing coatings to prevent the battery-like phenomena from occurring. The coatings insulate the pipe from the electrical charges and reduce the corrosive action. Metal pipe is subject to the worst types of corrosion but other pipe types are subject to corrosion as well. The biggest advantage of concrete and plastic drain pipes is that they are more resistant to corrosion than metal pipe types.

All types of pipes are subject to corrosion and all kinds of pipes are offered with corrosion -resistant insulation. A more detailed discussion of corrosion is covered in Chapter 5.

Coatings

Corrosion-resistant pipe coatings take several forms, which are chosen for their properties to resist corrosion. Coatings can be paints—including enamel and epoxy resin coatings; plastics—including Teflon and PVC; or common elements such as cement and coal tar. Almost all of these coatings are applied to steel pipes used for underground applications.

Most corrosion linings are added independently from the pipe manufacture. This means that to purchase new pipe or fittings that have a corrosion lining, the facility manager has to wait while the manufacturer fabricates the pipe, then ships it to a subcontractor who lines it with the corrosive resistant coating and then forwards it on the to facility.

PIPE JOINTS

In all pipe types, there is a decision to be made about the types of joints used between the different pieces of pipe. Almost anybody will recognize the joints used in residences, especially homeowners who have crawled under the sink to fix a drain. Joints are chosen for service and economy.

Screwed Joints

A screwed joint in pipe is made with a coupling that has been threaded at each end. Female threads in the coupling are "made up" with male ends screwed onto the ends of the female coupling. "Made up" is a plumbing piping expression meaning "connect to the next piece."

On screwed piping, the fittings are connected the same way. With a pipe system of entirely screwed fittings, it can be put together from one end to the other, but a repair means the pipe has to be cut and unscrewed in reverse sequence of assembly. This time-consuming process led to the development of other joint types. The advantage of a system with screwed joints is that tools to assemble the system consist of a pair of pipe wrenches. The joints make up quickly, experience few leaks and can be un-screwed and rescrewed repeatedly without leakage. Usually, Teflon tape or Teflon pipe dope is applied to the joint before as-sembly to assure a leak-tight joint and to make the joint easy to take apart when service is necessary For large-diameter pipe sys-tems, above six inches say, the costs of the fittings becomes expen-sive compared to other joint types.

Sweated or Brazed Joints

With the advent of copper pipe, holding the advantage of a thinner wall which made the pipe lighter and easier to fit up, the wall was not thick enough for cutting in the threads. Hence a coupling composed of the same material with a slightly larger diameter was developed. Copper pipe is inserted into the larger fitting and the joint is heated with a propane torch. When the joint reaches the correct temperature, solder is applied. The heat melts the solder and as the joint cools, the solder is drawn into the fit-ting, sealing it. The disadvantage of this type of joint has been getting the pipe hot enough without inadvertently setting the adjacent construction materials on fire. In addition, if there is water in the pipe, the water boils into steam and can pressurize the pipe beyond its strength. Water can also carry away the heat, preventing the joint from reaching the right temperature and resulting in the solder not being drawn completely around the joint. Hence, it leaks.

Brazed joints can be heated, pulled apart, sanded and reassembled without adverse affects.

There are several types of solder used for brazing joints. These solders have different melting temperatures and the correct type of solder is used to make up the right joints.

Glued Joints (Plastic Pipe)

PVC and ABS pipe are joined with glue. The glue is usually a special epoxy resin *made specifically for that pipe* and is applied to both the coupling and the pipe end. The two ends are pressed together and the glue forms a permanent bond between the pipe components. The disadvantage of this type of joint is that it cannot be used again. If the pipes are taken apart, the joint is ruined. The fittings of this type of pipe are inexpensive and are discarded and replaced with new ones when the joint is worked.

Bell and Spigot Joints

All types of pipe can be fabricated with a joint called a bell and spigot joint. The bell, so called from the bell shape at one end of a section of pipe, is fitted over the spigot end. A gasket is installed in the bell to prevent leaks. Bell and spigot pipe joints are used on pipe placed end-to-end and buried underground. The bell shape, when buried, prevents movement. There is a little flexibility in the joints of this type of pipe and long radius bends can be made by "pulling" the joints. That is, each joint is pulled out of line slightly from the preceding joint. Depending upon the bell, an angle of 5 degrees or less can be made between successive joints. For short radius bends in bell and spigot pipe a thrust block is poured. The thrust block, usually concrete, is placed between the outer wall of the pipe and the trench wall to prevent the pressure from pushing the two pipe joints apart when pressurized.

Mechanical Joints

There are several types of mechanical joints which are a combination of a bell and spigot. A ring on both ends have holes. Threaded rods run through the holes and the joint is held closed with nuts placed on the rods. Mechanical joints are effective beginning at the 6-inch pipe size up to 48-inch diameter size.

Compression Joints

In smaller sizes, up to about 4 inches in diameter, compression joints are used with fittings. Compression joints use a *union* to join two pieces of pipe together in a place where they can be taken apart and put together again. In the union, two plates press together using a threaded collar. The force of the threaded collar holds two pipes together with such force that there is no leak. The advantage of a compression fitting is that it can be taken apart quickly and reassembled many times. The largest pipe union is about 2-1/2 inches in diameter. Threaded joints, on the other hand, have a tendency to leak if taken apart and reassembled many times.

Flanged Joints

Flanged joints are used in mechanical rooms because they are readily disassembled and re-assembled during maintenance. Often, large valves are installed with flanged joints to facilitate quick removal and reassembly. A flanged joint is made up with a gasket between the flanges. The flanges can be screwed or welded onto the pipe. The flanges are expensive, however, and increase the costs of the installation. In general, flanged joints are easier to break and reassemble than mechanical joints. Large crescent, socket or open-ended wrenches are used to take apart these joints and there should be enough open space around the entire joint for free movement of the socket or crescent wrench used. The smallest flanged joint is for one-inch diameter pipe.

Welded Joints

Finally, joints in steel pipe are welded. The pipe is cut to length with a cutting torch and successive sections are welded in place. Sometimes, a machine is used to weld the joint while in other cases a welder—using a metal arc or metal-in-gas welding equipment—fuses the joints together. Welded joints are the strongest and are used for joining high-pressure steam and compressed air systems. Welded joints are also used for water systems. Usually, welded joints are fabricated on the jobsite and fitted up as they are welded together. Welded joints, obviously, are not intended to be disassembled. Many facilities use a combination of flanged joints and welded joints for their pipe systems.

Chapter 8
Pumps and Tanks

*P*umps *are used to push water through the pipes and tanks. Tanks provide a storage space for water. Tanks can store fresh drinking water, hot water or raw, untreated water. It is important that the tanks keep the separate types of waters apart from each other. Plumbing codes are designed so that pipe systems do just that.*

PUMPS

Almost all supply waters, and often wastes, have to be pumped. And as was previously stated about piping, many engineers make their entire living solely from the business of pumps. Pumps are designed for only one function—to move the fluid through pipe. For facility managers managing water systems, the most common type of pump is the centrifugal pump.

A centrifugal pump uses an electric motor to turn a shaft that has a blade. The blades (called impellers) push the water through them. Centrifugal pumps come in all sizes and shapes and are manufactured for numerous special applications. The size of the motor determines the amount of flow and the increase in water pressure.

All centrifugal pumps have a relationship between the pressure increase and the amount of flow. Example: As the pressure increases on a pump, the flow through it decreases. A simplified pump curve is shown in Figure 8-1. At a certain point, the pump cannot push the water higher and at another point, it cannot move any greater flow. These are the ends of the curve for that specific pump. Engineers select pumps that have the best efficiency between the ends of the curve, which saves the client money on energy costs.

A cutaway section of a pump is shown in Figure 8-2. The drawing shows the shaft, impeller and casing.

As with the previous discussion for pipe, there is a tremendous variety of centrifugal pumps constructed of a wide variety of materials. The materials affect the costs of pumps significantly. However, facility managers should be aware that the electricity purchased to drive a pump far exceeds the cost of the pump itself during its life. The facility manager is far ahead by spending slightly more for a more energy-efficient, low-maintenance pump than to try to save a few dollars on the initial purchase.

PIPE AND PUMP COST TRADE-OFFS

The question of energy costs raises an interesting point that exists between the costs of the piping and the costs of pumping.

Just imagine for a moment a big pump pushing water through a *tiny pipe*. The pipe is so small that it takes a very powerful pump to push the water through it to the users.

On the other hand, imagine for a moment a *large pipe*, being supplied with a small pump. In this case, the pump is quite undersized for the intended service.

Given these two extremes, the designer of a water system wants to optimize the investment in both pipes, pumps and electrical energy costs. Overall, the optimum combination of pipe size, pump cost, and energy determine the overall system costs.

To analyze the trade-offs between pipe costs and pumping costs, the designer needs to know 1) the facilities' electrical energy costs and 2) the expected life of the pipe system. The designer should already have an estimate of pipe and pump costs and often makes assumptions for the owner about energy and life. However, for a long-life project, the electrical energy costs affect the options more significantly than for a short life.

Given the state of modern mathematical modeling techniques and personal computers, it is possible for the facility planner to analyze many scenarios and select the most attractive option.

Utility construction costs are high because the cost of power for pumping is projected several years into the future. By increasing pipe diameter, energy costs are reduced because less energy is

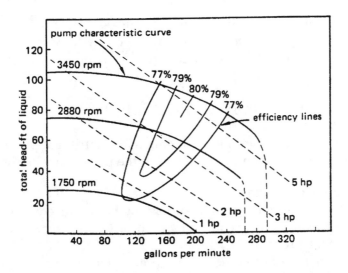

Figure 8-1. Pump performance curve. Courtesy: *Mechanical Engineering Reference Manual*, **Michael R. Lindburg, PE, Professional Publications, Inc., Belmont, CA, 1994.**

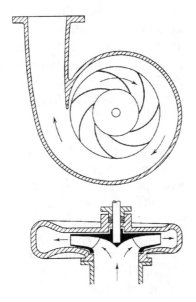

Figure 8-2. Cutaway view of a pump impeller. Reprinted from *Fluid Mechanics with Engineering Applications* **with permission from McGraw-Hill Book Co., New York City.**

needed to push water through large pipes than through small ones. Increasing the pipe diameter, while costly for large systems, will pay for itself with reduced energy costs over the system's life.

PUMP OPERATIONAL PROBLEMS

The two most common problems with centrifugal pumps are with the pump seals and with debris in the pump impeller blades.

Pump Seal Problems

Probably more frustration and agony has occurred by facility managers over pump seals than with any other water system management problems.

The shaft from the motor to the impeller must be separated from the fluid by a seal. The seal prevents water/liquid from leaking out of the pump. Thanks to many long hours spent by engineers and designers, today the design of seals for water pumps is excellent.

However, seals still have to be maintained and changed/repaired on a routine basis. When a pump seal fails, the pump has to be taken out of service for repair. If this happens when somebody in the facility wants to use water, there will be a significant number of complaints. To prevent these, most systems are designed with a pair of pumps that can alternate. When one pump is taken out of service for seal repairs, or any other reason, the other pump continues to provide the facility with the necessary water and pressure needed for operations. Depending upon the size of the pump and how easy it is for maintenance crews to work on it, a pump seal can be changed in an hour, or it can take an entire day. Some facilities keep a spare pump on hand and when maintenance is needed, the entire pump is pulled along with the motor, and a spare one is dropped into place. The original pump is serviced, and placed back into spares.

Many facilities standardize their pump sizes to reduce the need for spares throughout the complex.

Pump Plugging Problems

The other most frequent problem encountered with centrifugal pumps is one of plugging with debris, dirt, rocks or other material that either goes into the pump body and either is caught by the impeller blades or plugs the piping. A device is usually installed upstream of the pump to strain out this debris. Not surprisingly, it is called a strainer (see Figure 8-3).

A strainer consists of a wire basket that sits inside a bucket in the piping. The basket's tiny holes are sized to catch particles large enough to bind or harm the pump's impeller. The basket is set into a chamber and is removed for cleaning when plugged with debris. The strainer is usually isolated from the pipe with valves to pre-

Figure 8-3. A pipe strainer for a large swimming pool. The strainer protects the pump impeller and seals from debris.

vent draining the line when the basket is cleaned. The strainer can be bypassed for a short time while the basket is being changed or the system can use what is called a duplex strainer where one basket is working while the pump is running while the second is cleaned.

PUMP SUCTION
AND PRIMING

The two types of pump installations are called flooded suction and non-flooded suction. For the easiest maintenance, a pump with a non-flooded suction is best. This way, when the pump is off, the water runs back into the piping or reservoir, leaving the piping dry and reducing the mess that accompanies changing out pump parts.

A flooded suction is isolated from the reservoir or upstream piping with valves, but his type of installation will be full of water when the system is shut down, and provisions must be made to drain this water when pipe lines are pulled apart to remove the pump.

Pumps can become vapor locked when the water does not fill up the casing where the impeller is located. The pump turns, but water is not moved because the spaces are filled with air. Most pumps are self-priming and will suck the water into the impeller when they start. The key to self-priming is tight seals. In general, a leaky pump will not prime itself.

Quite often, the pump is designed to sit at floor level, just a few inches above the reservoir or the natural pressure of the water. These non-flooded suction pumps are close enough to the reservoir that they will self-prime and free drain. For maintenance, this is the ideal pump installation.

A pump that will not self prime can be primed by flooding the impeller with a hose or other source. Once running, it will maintain itself indefinitely. However, maintenance forces have to constantly re-prime the pump if the power fails and the pump shuts off.

For a complete discussion of maintenance of pumps, pipes and water systems, see Chapter 17.

TANKS

Just as water is pumped through pipes to where it needs to be, tanks are used to store it for later use. The most common tanks encountered by facility managers are hot water tanks. These tanks store up an adequate amount of hot water to be used by the facility when needed (see Chapter 12 for more on hot water systems).

In addition to hot water tanks, which are insulated with foam or fiberglass, elevated water storage tanks provide water for commercial and residential use. These large tanks are usually under the control of the water utility, but the facility manager may have to manage one or two. In some large buildings, tanks are installed to provide water supply near the point of use (see Figure 8-4).

Just as pumps are coordinated with pipe size in a matter of economics, tanks can be utilized in a water system to reduce the size of the pipe between two points. The tank then stores up water during periods of low demands and distribution pipes are supplied from the tanks when demands are high. In order to provide

Figure 8-4. An above-ground utility water tank. The small block-house encloses a fire water pump capable of pumping 1,200 gallons per minute.

enough pressure for the distribution system, tanks are often elevated. The purpose of elevating the tanks means that the water
does not have to be pumped after it leaves the tank. Water is
heavy and an above ground elevated tank is subject to large
weights. Design of elevated tanks is a specialty. One other problem with elevated tanks is their tendency to freeze if the water is
stagnant in them for an extended period during cold winter seasons.

Fresh water tanks can be installed in high rise buildings to
balance out pressures for a few floors.

In addition to elevated tanks, concrete or steel tanks can be
constructed below ground. This alternative is attractive in areas of
uneven terrain or hills. The water tank is located below ground
but on a hill so that it is still above the homes and businesses
where the water is distributed.

Concrete or steel tanks provide the most economical, attractive combination of life and cost. Many tanks are lined. That is the
inside of the tank is coated with epoxy or another inert material
to protect the system from rust or leaks.

Since large water tanks are critical to a city utility, codes for
tank design are rigorous and make the tanks relatively expensive.
The tank must be strong enough to hold the internal pressures of
the water. Tanks must be checked to confirm they meet seismic
requirements for the areas where they are constructed. In addition,
tanks have to have the associated piping coordinated to get the
water into the tanks and out again.

Finally, tanks need hatches for access to allow the tanks to be
inspected and ladders must be provided, if the tank is above
ground, to allow maintenance personnel to get to the hatches.
Many tanks have instruments that provide operators with the
temperatures and the water levels inside.

Chapter 9
Valves and Fixtures

*V*alves *are installed in pipes to control flows. Many types of valves are installed in pipe systems depending upon the need to regulate flow, maintain pressure or isolate components for maintenance.*

Fixtures are those plumbing elements that interface between people and water or wastewater systems. Fixtures include faucets, toilets and other water devices where people and water come into contact.

VALVES

Valves are used to control flows and to shut off flows from pipes. As with the previous discussions with pipe and pumps, there are many types of valves and their application is varied. Valves can also be operated by hand, with electricity to open, close and vary flows, or valves can be operated with air pressure (pneumatically). Air pressure operated valves work very well to control varying flows. These types of valves are usually installed in combination with a flow meter. The meter reads the flow and sends an electric signal to a device that controls the size of the opening of the valve. If more flow is required, changes allow the valve to open more; if less flow is needed, the valve is closed slightly.

Past use of computers and industrial application has caused tremendous growth in these types of valves and their servicing is not usually done with plumbers since electronics and pneumatics control the valves. A fairly common type of valve in a pipe system is called a solenoid valve. Solenoid refers to the device that controls the valve (generic term is actuator.) Solenoid-operated valves are either open or closed.

Hand valves are still used, of course. The three most common types of valves are still gate valves, globe valves, and ball valves (see Figures 9-1, 9-2, and 9-3).

Gate Valve

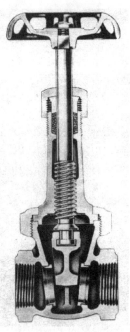

Figure 9-1. A gate-type valve. Reprinted from *Lyons Encyclopedia of Valves* with permission' Van Nostrand-Reinhold, 1975.

Figure 9-2. A globe-type valve. Reprinted from *Lyons Encyclopedia of Valves* with permission, Van Nostrand Reinhold Company, 1975

Figure 9-3. A ball-type valve Reprinted from Lyons *Encyclopedia of Valves* **with permission of Van Nostrand-Reinhold Company, New York City, 1975. Original illustration from Chemetron Corporation, Fluid Controls Division.**

The original valve is called a gate valve (see Figure 9-1). The gate valve is just that, a gate. With a series of threads on the stem of the gate, a handwheel is used to lower or raise the gate and increase or decrease the flow.

The problem with the gate valve is that for a low-flow application, the gate throttles off most of the flow. This will cause the valve to whistle or whine, indicating extreme wear on the gate. A lot of pressure will be lost when a gate is throttled to near zero flow.

Globe Valve

As a result of operational problems with gate valves, the globe valve was invented (see Figure 9-2). The globe valve, instead of a flat plate as the gate valve, is a round disk that sits flat on a round hole in the valve body. As the stem screws up and down, flow is even around the flat disk. Most garden hose and sink faucets are globe valves.

Globe valves and gate valves always seem to have a problem with leaking around the stem. Gate valves are especially notorious but these can usually be fixed in a half-hour or so.

Ball Valves

Finally, the favored application for most small systems is the ball valve. The ball valve is so called because a ball with a hole in it is used for flow control. When the valve is open, the hole in the ball is lined up with the pipe. To close, the ball valve the ball is turned so that the hole in the ball is crosswise with the pipe and there is no flow. Ball valves are not used for throttling; they are usually open or closed. Ball valves are favored by mechanics because they are very smooth-acting and easy to operate.

FIXTURES

Another major element of a water system is combined under the broad heading of fixtures. Fixtures for water and pipe systems are where the people come into contact with the water. Fixtures include all of bathroom and kitchen items along with a few special items. Fixtures include sinks, toilets, faucets, wash basins, slop sinks, urinals, bidets, bathtubs and showers. Common elements to most fixtures are that they are constructed of a non-absorbing material such as stainless steel or porcelain, hard plastic, stone, brass or bronze. The materials will not corrode in the wet environment, and are smooth for easy cleaning and sterilizing. People use fixtures and water to clean food, wash, bathe and eliminate.

One reason for successful health in advanced countries stems from the health benefits derived from the effective use of fixtures to prevent spreading of diseases between individuals. Bathrooms where people are able to wash and bathe prevents the spread of many contagious diseases common in non-industrialized nations.

Diseases such as typhoid and cholera spread rapidly through water systems and to the people who use contaminated waters. In this country, diseases such as hepatitis, cryptodispodiea and Legionnaires disease have contaminated many individuals. See Chapter 4 for an explanation of some of the impurities and pathogens in water supplies and methods of treatment. The bathroom fixture, with easily sterilized porcelain or enamel basins and bowls, combined with pipes to remove the waste, have prevented many diseases and have contributed significantly to people's health in our country.

In the 1890s, most of the common plumbing fixtures discussed here were invented. Their design has been refined over the past years until now they are quite effective, commonly available and inexpensive.

Sinks and Bathtubs

Sinks have a smooth finish, a drain at the bottom, holes for a faucet at or near the top, and an overflow that drains the sink before spilling over. Sinks also come as slop sinks, a deep recessed sink placed closer to the floor than the standard sink height of 42 inches. Mop sinks are usually right on the floor in the corner of a janitorial closet. The janitorial staff can rinse mops and mop buckets without having to lift them off the floor. These sinks are combined with a floor drain. (See Figure 9-4 for a manufacturer's standard catalog mop sink product).

Like sinks, bathtubs come in all sizes and shapes. However, facility managers do not often deal with bathtub problems since a bath is more of a residential item. However, since the late 1970s, the jetted tub or spa has become a modern attraction at health clubs, gymnasiums, hotels and some corporate fitness centers at large facilities. The jetted tub is more like a spa than a bathtub, although they are similar in appearance. Spas are complex items combining pumps, jets, controls, drains and are not really fixtures but fall into another category (see Chapter 14).

Water Closets (Toilets)

It is unfortunate that the water closet/toilet is such a personal item, because it is not really celebrated for the truly marvelous invention that it is. The modern toilet with its smooth bowl to prevent sticking of wastes; rim flush to moisten and carry away solids and paper; and self-priming siphon to eliminate all of the material in a single flush is a tremendous improvement over its predecessor, the chamber pot, which was emptied out the

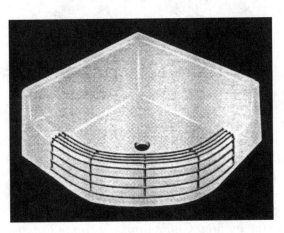

Figure 9-4. A mop sink. Courtesy: American Standard, Inc.

back step every morning for several hundred years before. The fact that toilet design has changed very little since its invention over 100 years ago is evidence of its simple, effective design.

Recent attempts at water conservation have generated regulations to reduce the amount of water used per flush. The old standard was 3 gallons per minute flow. New U.S. EPA standards for water-saving efficiency has reduced this flow to 1.6 gallons per minute. However, concern exists within the industry that reducing the volume per flush has negative benefits in draining, carrying solids in flows and increased deposits inside piping. In addition, the fluid flowing to the treatment plant contains more solids compared to the water and affects the effective operation of the plants. See Figure 9-5 for a cutaway view of a pictorial description of how a water closet works.

Water closets can be either wall-mount (totally mounted to the wall) or floor-mount (totally mounted to the floor.) The great advantage to the wall-mount is it is easier to plumb and the bathroom is easier to clean by maintenance personnel. The floor behind a floor-mount toilet in a public restroom is often an

1.   Tank ball letting water into bowl, starting the flushing action.

2.   Tank ball following the water out of tank.

3.   Tank ball closing off water draining into bowl. This completes the flushing action in the toilet.

Figure 9-5. Cutaway view of a water closet showing how a toilet works. Reprinted from *Step by Step Guide Book on Home Plumbing* with permission of Step By Step Guide Book Co., West Valley City, Utah.

unsanitary place subject to scrutiny by health regulators.

The facility manager will be careful to choose the low-waterflow water closets when remodeling. However, the design should be coordinated with the plumbing to assure the flow of water is fast enough to carry away the wastes.

Urinals

Urinals perform many of the same functions as water closets. Essentially, they are for men only. As with water closets, urinals are smooth and completely rinse the inside surface each time to prevent buildup and allow disease to spread. Urinals are mounted onto the wall. In the past, urinals were recessed into the floor slightly, but these types have not been installed in new construction for many years because of the difficulty of plumbing them while the concrete for the floor was placed in addition to the inability to service them once the concrete was cast in place.

Water flows for urinals are less than for water closets. The new standard for low-flow urinals is 1 gallon per minute.

Bidets

Not too common in the United States, the bidet is more common to European hotels. Similar to a toilet, the bidet is used for washing the genital areas. It can be either floor- or wall-mounted, although the models are usually floor-mounted. Plumbing faucets are provided for fresh clean water for washing (see Figure 9-6).

Figure 9-6. A bidet, not common in the United States, is often found in European hotels. Courtesy: American Standard, Inc.

Showers

Showers are provided as bathing facilities in exercise areas, plants where workers change after shift work, in hotels, schools and in public bathing/beach areas.

Showers are provided with a faucet with a nozzle for controlling the flow. The shower area is usually separated from drying areas with a curb. A slight rise or a sloped floor greater than 1/8-inch in one foot will be too steep and bathers may trip.

The floors of showers are usually ceramic tile with a slip-resistant finish. Some old plants used wood slats called duckboards. These duckboards provided a breeding ground for bacteria underneath and have since been banned by most public health agencies. Showers have floor drains to carry away the water.

Shower fixtures in recent years have become sophisticated with some unique pedestal types that are free standing from the center of the floor. A drain is also designed into the foot to catch wastewater (see Figure 9-7).

A facility manager should be cautious and allow plenty of spacing between shower heads, as the bathers waving arms and bending over do not wish to bump into other bathers. Public health regulations require separate shower facilities for men and for women at public facilities.

Figure 9-7. A center pedestal shower fixture, typical for what is installed in large dressing rooms

COMBINING SYSTEMS

Now that we have a basic concept of the water system's components, are ready to take the next step toward understanding how the overall water and piping system works. The individual components are combined into an integrated system. Proper system management requires knowledge of how the individual components work together.

Chapter 10

Instrumentation, Hydraulics, Plumbing

N ow that the fundamental components of a water system have been examined, a brief discussion of some of the science behind instrumentation is the next step to understanding the system in order to manage it effectively. Meters and gauges tell us how well the hydraulics systems are performing. And systems must be installed to meet certain standards that are the results of the flows, pressures and fluid properties.

INSTRUMENTATION

A good facility manager needs to receive constant information about how the water system is operating before it can be successfully managed. In order to provide this information and make it usable, *instruments* are used to provide managers with the information necessary to keep the system working in safe, efficient order. Instruments that provide this information include flow meters, pressure gauges and thermometers.

Flow Meters
Flow meters are installed in water systems to record the amount of water used. All flow meters work on the basic principles of fluid mechanics, some of which will be shared briefly in this chapter.

Water flow is measured with several types of instruments that provide varying degrees of accuracy depending upon the cost. Some meters are inexpensive and the more accurate the meters are the more costly. It has been said, "Water is cheap." The point is that people should not have to pay for more elaborate systems than are necessary. The facility manager should not have to spend several thousand dollars for a flow meter when the costs of the water he is measuring are low. This philosophy is true in the areas of instrumentation and especially so in the application of flow meters. The least expensive instruments should be used, provided they are accurate enough to do the job adequately.

Basic Flow Metering Concepts

There are basically two types of economical flow meters for water systems. Each is chosen for cost, ease of application and accuracy. In the science of flow measurement, a meter causes a change in the shape of the flowing liquid. This change in shape causes a change in the pressures which is measured using pressure gauges. The changes in pressure and the known shapes are used to calculate the *velocity* of the fluid inside the pipe. Then the *area* of the flow is calculated as well as the volume, volume being the velocity of the liquid multiplied by the area.

In these calculations, the areas are fixed but as the flow increases or decreases, the pressures change. A meter is designed to change the shape of the flow significantly—otherwise the pressure readings do not fluctuate enough to allow the velocity to be accurately measured.

In the English System of units, it takes a bit of math to convert the pressures and inches into gallons but this is the quick and simplified version of how velocity meters indicate flows. This type of metering is one of the oldest and has proven reliable and practical for many years. These types of meters are called orifice meters, venturi meters, or Pitot tube meters. Figure 10-1, 10-2 and 10-3 show these meters respectively.

Another way to estimate volumes from the velocity is to insert a small propeller in the fluid stream. As the water passes over the propeller, the velocity is calculated. The method here was to build a propeller of a standard shape and, in a laboratory, push water through the pipe at a known speed. Using small gears, the turning propeller turned a dial that was hooked to a clock face or to an odometer face similar in many ways to one on an automobile. By reading the numbers on the dial, the volume of the flow was estimated. This type of meter

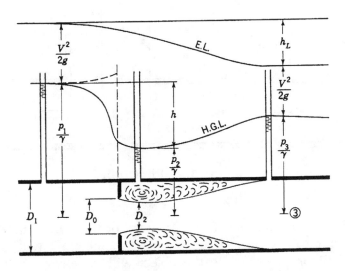

Figure 10-1. Orifice meter, in a pipe.

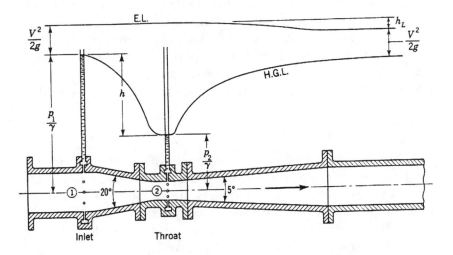

Figure 10-2. Venturi meter,
with an entrance shaped like a cone.

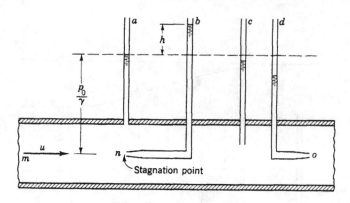

Figure 10-3 Pitot tube.

is called a *turbine meter*. The advantage of the orifice meter over the turbine meter was that an object in the flow will damage the small propeller while the object will usually pass through an orifice or venturi-type meter.

The third type of common meter utilized in facility water systems is the disk meter. A disk meter more accurately measures flows by using a flat disk that sits in the meter housing on an angle to the flow. As the water passes through the meter, the disk turns, similar to the propeller. The turning disk is geared to a clock face and the gears are designed to provide numbers that correspond directly to the number of gallons used.

Weirs

For outside water systems, a couple of types of devices are used to measure water flow in a canal or other open channel. Some of these types of water meters are installed in wastewater systems at treatment plants or in water lines inside manholes where flows can be measured. Flumes and weirs, as typical instruments, change the shape of the flowing water. The area is known from the depth and the shape of the weir. Most of these types of flow meters come in standard sizes with the flows computed based on depths. Typical weirs include the V-notch weir, Cipoletti weir and the box weir. Figure 10-4 Shows a typical box weir.

Figure 10-4. A weir with a concrete weir structure. Courtesy: U.S. Department of the Interior, Bureau of Reclamation.

Parshall Flumes

Similar to weirs, but more accurate in flow measurement and more effective over a broad range of flows, is the Parshall flume. Like weirs, Parshall flumes are used to measure open channel flows in ditches or other irrigation or wastewater facilities.

Parshall flumes for small flows can be purchased from irrigation companies made from galvanized metal. They are lifted into place. For large flows, Parshall flumes are constructed of concrete. Figure 10-5 shows a typical Parshall flume in a canal.

Figure 10-5. Parshall flumes. Courtesy: U.S. Department of the Interior, Bureau of Reclamation.

Electronic Metering

In the modern era of the computer chip, the pressure or depth is converted into an electric signal. The signal is used to carry forward all the high math, then translate the information into a direct reading of flow. At the push of a button, given the recording and processing capability of the computer chip, the flow can be read in gallons per minute, cubic feet per second, liters per minute, gallons per day and so on.

In the different industries connected with utility services, the units are all measuring a volume but the terms used are not all the same. For large sewage treatment plants, the units are discussed in *millions of gallons per day*, or MGD. For reservoirs and lakes which are much larger, services are discussed in terms of *acre-feet per day*, or per month or year. For monthly billing for a customer, the facility might want to charge by simply the gallon. However, what the bill really means is gallons per month.

Throughout the day, week, month or year, the facility manager's flows are going to vary. Hence a totalizing flow meter is used to plot the highs and lows. These can be recorded using other instruments or strip charts. Another type records flows on a flat disk, essentially a clock face that turns and the pen moves out from the inner circle depending upon the flow changes. These are called circle chart recorders.

More Sophisticated Meters

Other types of meters measure the velocity using ultrasonic sound, coriolis effect, vibration and magnetic force. Each of these types are more expensive and they have advantages and disadvantages in terms of price, accuracy, installation and service ability.

The most significant recent new trend in flow metering has been to install computers integral with the meter that includes a modem. The computer modem dials the home office every day and downloads the daily readings where it is tallied. A computer hooked up at the main office receives the data from the meters. Computer programs automatically print the bills to be mailed out to the clients.

Flow Meter Installation

Flow meters sometimes fail. Either the device becomes

plugged, the taps where the pressures are being read foul and do not read correctly, the case cracks, the gauges stop reading or the gears strip. While these types of events do not happen often, it may be necessary to change the *meter*, take it apart and repair it, or replace it.

Facility managers should be sensitive to the problems of shutting down the water lines or wastewater lines to service the meter. Small meters, those installed in pressure lines 6 inches in diameter or less, can be installed within a few minutes. A large meter, say in a 48-inch diameter line, could take a couple of days to replace depending upon how long it takes to drain the lines and where the drained water will be sent.

A *bypass line* is often installed around the meter to allow flows to pass through the system when the meter is taken out of service for repair. This way, the water service continues, the line with the meter in it is shut down, and the meter can be removed and repaired or replaced. The facility should remember to bill for water supplied while the meter is out of service. Past billing records are used to estimate water supplied while the meter is out of operation.

Pressure Recorders

Instrumentation and recording of pressures is the second most used of the water system measuring tools. (There are usually more pressure gauges, but they do not lead to as many arguments.) *Pressure gauges* measure the pressures of the fluids inside the pipe. Since water will flow from regions of higher to lower pressures if line water is flowing, plumbers and maintenance personnel use pressure gauges to tell them the direction of flow.

The other item of information from pressure gauges is the amount of stress on the inside walls of the pipes. If pressures are too high, risk of pipe rupture is a concern, while if pressures are too low, risk of contamination of the pipe from waste lines is a concern.

Most pressure gauges indicate locally, however, electric *signals* can be used to send information to a remote location many miles away, like flow meters. Pressure gauges are often installed with a valve so that the valve can be closed, the gauge removed and replaced and then opened again. See Figure 10-6 to view a typical pressure gauge.

A pressure gauge is usually tapped into the side of the pipe with a small hole, say 3/4-inch to 1/4-inch.

A direct-reading gauge has a fine spring inside, with a small flexible diaphragm. As the pressure increases, the diaphragm moves. The diaphragm is attached to a small mechanical linkage that in turn moves a dial on a clock face. Gauge faces come in all types. A large gauge, perhaps 4 inches in diameter, might be marked every 2 pounds while other gauges will be marked in 5, 10, or 25 pound increments.

Remote-reading gauges are installed similarly to direct-reading gauges. Remote-reading-type gauges use the same diaphragm expansion concept, except that the mechanical link to move the dial on the clock face is replaced with a dial on a rheostat (an electrical device.) The rheostat measures a varying electric current, then computer chips and other electronic components converts the electrical signal into a digital readout similar to what is seen on an electronic calculator. Again, this signal can the sent through wires to a remote station many miles from the meter.

It has been shown that it is more likely that the dial-type meter will be misread than the digital meter, but for most installations this reading may not be that important. The problem with dial-type pressure meters is that personnel will sometimes not

Figure 10-6. A typical pressure gauge. Pressure gauges measure the pressure of the fluids inside the pipe. If pressure gets too high, the cause should be determined and corrected immediately.

write down the correct reading because a mistake is made interpreting the position of the needle.

Like flow meters, pressure instruments and pressure recorders can become expensive. A dial-type pressure gauge might be $25.00 while a digital device will be more expensive. In addition, the electrical devices require power from an electric source. Often, a design specifies an electric-type meter or instrument when the meter vault is in the middle of a large field or along the edge of the highway and there is no electrical power available. The cost to run conduit and wire to the metering vault from one of the nearby buildings can exceed the cost of the meter itself. Facility managers should not be put into a position to purchase any more costly pressure gauges than are necessary to accomplish the mission.

In addition, the facility manager should note that electronic instruments require a different staff skill level. Staff to repair and purchase lots of electronic instruments can result in the need for hiring additional staff that may have not been initially planned.

Pegging Pressure Meters

A word of advice for facility managers about pressure gauges: occasionally, water hammer or mismanagement of the system will result in a sudden extreme pressure increase. If the increase is more than can be read by the pressure gauge, the increase will cause the needle of the gauge to hit the stop at the extreme end of its range. When the needle hits, it will give a little metallic sound, most accurately described as *tink*. This is called "pegging the gauge," and it is a well known term to plumbers and engineers familiar with this type of equipment and systems. "Pegging the gauge" is not good—in fact, it can be dangerous because it indicates extreme pressures that could be serious. Any craftsman who does not treat "pegging a gauge" seriously should be disciplined. If they continue, they should be replaced.

Pipe supply systems can withstand a small measure of sudden extreme spikes, but these are common to start-up. The cause of the spike should be identified and the system modified immediately to prevent it. Most systems will "peg" several times before they fail, but if they do "peg" and the "pegging" isn't stopped, the pipe system will fail and a potential disaster is about to happen.

Thermometers

Monitoring temperature is important where water is used for heating and cooling. Thermometers are used to measure the temperature of hot water systems.

Temperature readings can also be important at sewage treatment plants were microbes are used as a part of the wastewater treatment process. Other water supplies may need the temperature monitored because of processes downstream where the water temperature is important—cooking or food processing, for example.

The simplest way to test hot water is to turn on the hot water tap and stick a thermometer in the flow. Otherwise, a hot water tank can have a thermometer on it to tell the operations personnel what the temperature is inside the tank. Most dial thermometers use a diaphragm or an expanding bellows that changes shape with the increase in temperature. The diaphragm is attached to a spring that moves a dial on a clock face. Other thermometers operate on a bulb principle. For these, a recess is attached to the hot water tank and the thermometer bulb inserted into the recess. The recess in the tank or pipe is called a *thermowell*.

Like the flow meter and pressure gauge, thermometers can be hooked up to digital readouts to give the temperature in direct readings. The advantages are the same as pressure gauges in writing down the readings.

Red-line/Warning

Gauges can be purchased with red and green backgrounds on the clock face.

Green means, of course, safe, and red means danger.

The advantage of this type of dial face on the gauges is that the operator does not have to understand what pressure or temperature is too high—the gauge's red line notifies him immediately.

CALIBRATION

All instruments should be checked regularly to verify their accuracy. A meter or gauge that does not read accurately is not worth having in the system and should be removed or replaced. Most facilities have a calibration procedure where instruments are taken out and checked on a routine basis.

For complex industrial plants and regulated industries such as hospitals, routine calibration is a requirement and the facility manager is asked to produce the records that calibrations and readings have been performed. This investigation is called an audit. It is not an audit of the books the see where the money went, but instead a check of the procedures used in the facility to confirm that risk is minimized via routine checks.

HYDRAULICS

As mentioned above, there is an occasional need for a facility manager to understand the rudiments of hydraulics to enable him or her to have a better understanding of the problems the staff and engineering consultants face in managing a water system. The very basic simple terms used in fluid mechanics are included here.

Water flows to the lowest point, water always runs downhill, and water seeks its own level are all fundamental non-scientific explanations of water flow. The specific term for understanding water flow is called fluid mechanics.

Water has weight but it will not compress. Suppose we have a container full of water. Squeezing water in one place will cause all the water in the container to be squeezed the same amount. If squeezed hard enough, the weak wall of the container will break and a leak will result.

One of the most significant principles in hydraulics is gravity. Water will run downhill until something contains it, then it will fill the space, flowing into and filling the shape of the container from the lowest point upward until one of three conditions are met:

1.   The water stops coming in and the fluid stops moving and fills the space.

2.   Water keeps coming in until the space is full and then water will flow out into the lowest adjacent space and begin to fill it until it is also full.

3.   If an opening is created below the water surface in the space, the water will flow out of it.

If water is flowing in faster than it is flowing out, the volume in that space will increase. If the water flows out faster than it is flowing in, the volume of water in the space will decrease.

Pipe, by providing a uniform round surface follows these principles by filling from the bottom up and filling the space. However, pipe in a upside-down U shape will have a bubble in it, as flow fills the shape up on one side it will flow over to the other side until each side of the U is filled. A bubble of air will be present at the top of the upside-down U.

This bubble can be flushed if the flow is suddenly increased and decreased but it may not be completely eliminated. The bubble of air pinches off some of the flow. As a result, many systems will have an air vent to let the air out of the pipes at high points. In the reverse case, when the pipe is full and water is drawn down, there needs to be a way for air to enter through the vent.

For large utility pipes of high diameters and thin steel walls, the pipes have been sucked flat similar to a paper soda straw. In the early 1950s, in a steel pipe siphon over the Ogden River Canyon in Utah, a pipe was sucked closed by this phenomena. The engineers looked at it and thanks the ductility of steel, a decision was made to fill the pipe again. It worked. Air vents were installed the next day.

Fortunately for facility managers of most systems, the common steel pipes discussed in Chapter 3 will not have this problem because the pipe wall is thick enough that the suction of the water does not have enough force to collapse the pipe.

Venting pipe supply lines at high points to release these "bubbles" is an important element of any pipe system.

Water Hammer

Water hammer is the sudden pressure spike when a flowing fluid is suddenly closed off. It is called water hammer because of the rumble and shudder of the piping as the moving fluid slams to a sudden stop. What happens here is that a long tube of water is moving steadily through the pipe. The weight of this water can be extensive and while the pipe is not moving, the water flowing inside the pipe is. Essentially a lot of weight is moving fairly fast and suddenly, when the valve closes, the water stops. In theory, the spike of an instant closure is infinite. In reality, the valve does not close in an instant and the water and pipe stretch slightly. The rattle is usually the result of movement in the pipe as it stretches. The sound of water hammer is indicative of pressures great enough to bend metal and continued operation of a pipe system with a water hammer problem can potentially destroy the pipe.

Methods to prevent water hammer include the use of a stem rise of pipe which forms a dead column of air just upstream from the valve. This short piece of pipe becomes a bubble of air that forms a "cushion" for the water when it suddenly stops.

Stem risers used for residential and small pipe systems in the 1/2- to 3/4-inch category are usually installed behind the sink or lavatory. For preventing water hammer in larger pipes, valves are selected which close slowly using a screw mechanism. This gradually slows the water flow down before closing it off completely.

Case Study: Water Cost Savings Can Pay

Often, supply water is used for process cooling, but it is best to recycle this water and treat it.

One facility used water to cool a compressor that ran seven days a week. The water wasted over a year would have paid for a complete refrigeration recirculation system, had an economic analysis been done.

When an economic analysis finally was done, the recirculation system was installed the following week, saving water bills and preventing a needless waste of fresh, treated water.

For large utility systems, the valves are electrically timed to close at a rate that allows the entire tube of fluid to slowly come to a halt. Engineering studies and computer programs are used to model the potential of water hammer for large systems. This type is analysis is not expensive, and it is well worth the evaluation for any pipes larger than about 18 inches.

More Complex Hydraulics

Without going into a lot of mathematics and complex terms, this short discussion of fluid mechanics may prove beneficial to the facility manager when discussing water management problems with engineers or maintenance personnel.

A pipe system of flowing water has three elements of energy: the height of the pipe, the pressure inside the pipe, and the energy of the velocity as it flows (see Figure 10-7). The height of the pipe is relevant to the fluid upstream or downstream. The pipe wall must be strong enough to keep the pressure inside the pipe from bursting through the pipe wall. For the right pipe to be selected, the pressures need to be calculated. For building systems, codes have standardized the pressures and the pipe types to prevent rupture. For utility systems and aqueducts, engineers prepare a detailed analysis to ensure recommended pipe pressures are not exceeded.

To conclude our explanation of hydraulics, we also need to understand friction losses. As the water flows through the pipe, the inside wall of the pipe is not moving and the water right at the wall rubs against the pipe and it slows down. Water out in the

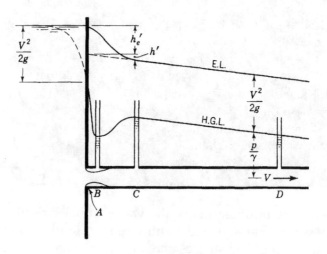

Figure 10-7. Engineering drawing illustrating hydraulic relationships.

center of the pipe moves faster because it does not get affected as much by rubbing against the walls. The faster the water moves, the more friction builds up at the walls resisting the flow. Over a long stretch of pipe, this friction loss is calculated for different velocities. As the water flows, the friction reduces the pressure. Pumping systems must push the water through the pipe and overcome the drag of the water rubbing along the walls. Most of the engineering calculations for water systems are to account for these friction losses. Some pipes have inside walls that are more slippery than others.

To account for friction loss, engineers calculate the pressures from elevations and flows. The engineers then select the *next standard size* of pipe that meets these requirements. It is not practical to try to have an odd size fabricated for just one job. Finally, calculations are made at one-third capacity and at two-thirds capacity. Flows are not always at the maximum, and it is necessary to know what the pressures and velocities will be when flow is less than the maximum amounts.

Today, most engineering companies use standard computerized mathematical models that run on personal computers to solve these problems. The software costs $150-$5,000 depending upon the size of the system the computer can handle and the number of flow and pressure variables.

Smooth Means More Water

In a recent advertisement in a plumbing magazine, a manufacturer boosted that his pipe had a better friction coefficient. What this translated to was that the manufacturer's pipe could carry 10 percent more water than its competitors because the pipe's smoother insides did not offer as much friction loss to the flow of fluid through it.

These computer programs allow quick calculation of alternative methods and pipe sizes; modeling has proven effective at optimizing systems.

The theories used by engineers to estimate pressures, flows, and sizes is based upon experimental methods, tests and results obtained early in this century. They are tools for estimating how new systems will behave, but they are not exact. Built into the formulas is a slight "factor of safety," such that flows *should* be slightly more and pressures slightly less. Depending upon the details of calculations, the attention paid to construction, and the accuracy of assumptions, most large supply pipelines deliver within 2-5 percent of the design flows. Table 10-1 provides the names and vendors of some water system computer modeling software programs. The facility manager should ask the engineer about the various options examined for his system and make sure the consultants have considered most of the significant elements. The facility manager should also request an assessment of both the tangible and intangible assets of the options.

Table 10-1. Computer modeling software program vendors.

Program	Vendor
The Crane Company Fluid Flow Software	The Crane Company
Plumbware's Plumbcad Software	Plumbware Company
Hydraulic Calculation Program	Custom House Designs
AutoCad Plumbing	AutoCad Software
Flo-Series, Integrated Piping Design	Engineered Software, Inc.
	1015 10 Ave., SE
	Olympic, WA 98507
	(800) 786-8545
Cybernet Version, Autocad Integrated	Haestad Methods
	37 Brookside Road
	Waterbury, CT 06708
	(800) 727-6555
Design & Estimating Software	Elite Software
	P.O. Box Drawer 1194
	Bryan, TX 77806
	(800) 648-9523

Wastewater Hydraulics

Wastewater pipes carrying sanitary sewer and storm water are almost always larger than supply water pipes. The larger diameter is the result of two factors the facility manager needs to understand if he is to be a successful water manager. The first of the two factors is that the pipes carry solid material. For sanitary sewers, the solids are bits of paper, garbage and human excrement. For stormwater, the solids are paper, trash, stones, sticks and other debris. The second of the factors that cause wastewater pipes to be larger than supply water pipes is a result of the hydraulic design.

How to Quickly Estimate Pipe Flow Capacities

Most engineers, use the velocity to select the pipe size using what is called the continuity equation. An engineer estimates a first guess velocity of 10 feet per second and calculates a size (diameter) from that number. For a facility manager, all he needs to do is look at the size (diameter) and multiply the area of the pipe by 10 feet per second. This will be close to the maximum flow that can be carried by the pipe. This trick is the method used in fire-fighting codes for firefighters to estimate pipe flow capacities. But the facility needs to be careful because the engineer sometimes uses different velocities as a result of pressure or friction losses. The typical 10 feet per second for pipe sizes 3 inches in diameter and larger is the fastest "safe" velocity. Faster velocities con be used but bubbles, bends and friction losses take a toll on these rule-of-thumb numbers.

Wastewater systems are designed to flow less than full. That is, the flow of the water combined with solids does not completely fill the pipe. In theory, the ideal depth for a round pipe that does not flow full will be about 8/10 of the diameter, but almost all sewer piping is oversized so that it does not flow at the ideal depth. In sanitary sewers and to some extent in stormwater sewers, organic matter in the water is broken down by the action of microbes. The process releases water and gases. The two most common gases released are carbon dioxide and methane. In addition, hydrogen sulfide, an extremely foul-smelling gas, is a by-product of the organic decomposition process. The blanket of air lying on top of the water inside sewer pipes provides a channel for this gas to escape (see Chapter 4).

Sewer lines also use the blanket of air for "floating" debris that would otherwise rub against the underside top of the sewer pipe and slow down or plug up the flow.

Further, sewer lines flowing partially full are not pressurized. If there is a leak in the piping on the top half of the circle of pipe, leaks will allow water in, instead of being forced out. For this reason, sewer pipes are buried in the ground below the level of the supply water lines.

Sewer lines have to slope downward at about the same angle that the water will flow. If the sewer line is too flat, it will fill up with water and become pressurized. If it is too steep, all the water will rush out from under the solids, and the solid material will be left to decompose in the pipe instead of being carried downstream to the sewage treatment plant.

Without calculations, plumbing codes require sewer lines to slope downward at 1/4-inch of fall per foot of horizontal run. This works out to be one inch in 4 horizontal feet, or 2-1/2 inches in 10 horizontal feet. This rule, by code, combined with the sizing requirements for the sewer pipe diameter, has proved satisfactory for the past 90 years. It is backed up by years of research and hundreds of technical papers.

If there is an obstruction in a building that is in the way of the sewer pipe—most often, it is a beam that holds up the floor—either the sewer pipe or the beam must be moved. Raising the sewer pipe is difficult because it means the toilets have to be moved or raised. If the pipe is lowered, it can mean the sewer pipes must run along the ceiling of the floor below.

The third alternative is to cut or move the building's beam, but this can be a problem since it may weaken the structure enough the building could wind up in danger of collapse.

On a big construction job, ironworkers who install the beams love to squabble with the pipe fitters, who install the sewer pipes over this issue. It usually requires some redesign to make the pieces fit together for which the owner, if he is not careful, ends up paying.

Outside buildings, sewer pipes have the slopes calculated by engineers. For any flow and pipe diameter, there is an ideal slope to make the sewer water flow at the proper depth. The larger the pipe, the flatter the slope. The calculations to figure these depths, flows and slopes are slightly complex, but since the science has been fairly well established, computer programs and piping handbooks define the ideal slopes, diameters and flows. In the same way pipes are sized for supply lines, pipe sewer designers choose the *next larger pipe diameter* for sizing sewer pipes.

For utility work, a sewer manhole is placed every 300 ft. or so along the sewer line. The exact distance is often specified by the city or local jurisdiction. Manholes allow utility operators to measure the flows and to unplug the lines. Sewer lines are always straight between the manholes, and manholes are where the directions and slopes of sewer lines are changed. Flows can come into manholes from different directions and all flow out one large single line. If a line is plugged between two manholes, the upstream manhole can be pumped, to keep sewer lines flowing while crews work to unplug the affected line.

A couple of other items relative to sewer systems will complete this section of discussion about sewer and wastewater piping. Occasionally, sewer water is too low or the ground is too flat to achieve the necessary slopes for sewer piping. When this happens, a "lift station" is installed. A sewer lift station uses pumps to lift the sewage water up, usually only a few feet so that it can run downhill again. For long flat lines, a series of lift stations are used. Pumps are used to lift the water, but because sewer water has solids in it, a grinder is installed upstream of the pump. The grinder breaks up the solids in the flow so that it can be pumped by the lift pump. Some special types of pumps are also used that can lift the water and the solids. Pumps of the peristolic or diaphragm type are sometimes used.

Problems with lift stations and grinders come when the pumps fail, because the sewage flows continue and cleaning up sewage at a failed lift station is a mess. For this reason, almost all lift stations are fully doubled—double pump, double grinder, double power, backup power supply, etc., in order to reduce the chance of failure and to facilitate easier cleanup.

PLUMBING

Plumbing is the process of installing or fixing pipes inside buildings. The word is also used to mean the pipe system itself. In ancient Roman times, pipes and drains were made from lead and a lead worker was a pipe worker. The word "plumb" originally meant "lead" in Latin. In the 18th century, pipes were also made from lead and even 50 years ago, pipes were sealed with molten lead. Hence the trade name for a worker who installed or repaired pipes became "plumber."

Installing pipe inside buildings requires knowing and understanding water flow in addition to understanding pipes and how they fit together. Plumbing is a skilled trade and plumbers spend years learning the codes. In many states, a plumber is apprenticed, then takes a test before he is licensed.

Within buildings, supply and waste pipes are installed according to laws designed to protect the building's occupants. Wastewater potentially carries disease and supply water can be contaminated if the two systems are installed incorrectly.

Supply pipe installation has already been discussed previously in this chapter. There remains a short discussion about wastewater piping for the facility manager to become fairly knowledgeable about plumbing systems.

Traps and Vents

Plumbing drain systems use traps and vents to "trap" the sewer gases and "vent" them to the outside of the building. A figure of a P-trap is shown in Figure 10-8. The trap fills with water and keeps the sewer gases from backing up into the room. A vent is installed downstream from the trap to let the gases out of the building. A sketch of a trap and vent is shown in Figure 10-9.

Traps and vents are required by the codes to be within fixed distances from the drain (see Chapter 3 on regulations). If they are too far from the drain, the effect of the trap is not realized. A toilet, by the way, is a sophisticated and special type of trap.

Supply and waste lines inside buildings are sized according to the size of the drains and the numbers of fixtures within the buildings. The sizes are also regulated by the plumbing codes. Table 10-2 provides a sampling of some of the sizing requirements from the Uniform Plumbing Code.

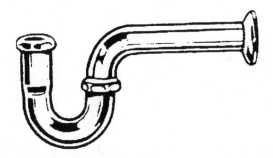

Figure 10-8. A plumber's P-type trap fills with water and keeps sewer gases from backing up into a room. Courtesy: Step By Step Guidebooks, Inc.

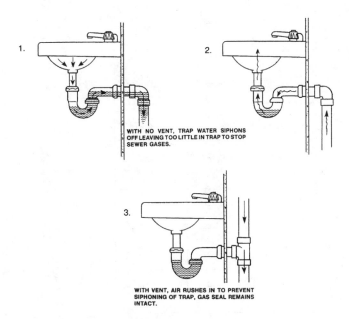

Figure 10-9. A trap and vent system. Gravity brings water through the pipe, causing a vacuum in the pipe above it, called siphon action. The suction effect draws the water from the trap nearly completely, exposing the fixture to easier transmission of sewer gases through the pipe. Venting prevents this suction, drawing gases from the outside air, typically through roof vents. Reprinted from *Step by Step Guide Book on Home Plumbing* with permission of Step By Step Guide Book Co.

Table 10-2. Drain Pipe Sizing from the Uniform Plumbing Code. Reproduced from the 2000 edition of the Uniform Plumbing Code™, copyright 1999, with permission of the publishers, the International Association of Plumbing and Mechanical Officials. All rights reserved.

Size of Pipe, Inches (mm)	1-1/4 (32)	1-1/2 (40)	2 (50)	2-1/2 (65)	3 (80)	4 (100)	5 (125)	6 (150)	8 (200)	10 (250)	12 (300)
Maximum Units Drainage Piping[1]											
Vertical	1	2[2]	16[3]	32[3]	48[4]	256[6]	600	1380	3600	5600	8400
Horizontal	1	1	8[3]	14[3]	35[5]	216[5]	428[5]	720[5]	1 2640[5]	4680[5]	8200[5]
Maximum Length Drainage Piping											
Vertical, feet (m)	45 (14)	65 (20)	85 (26)	148 (45)	212 (65)	300 (91)	390 (119)	510 (155)	750 (228)		
Horizontal (Unlimited)											
Vent Piping (See note) Horizontal and Vertical											
Maximum Units	1	8[3]	24	48	84	256	600	1380	3600		
Maximum Lengths, feet (m)	45 (14)	60 (18)	120 (37)	180 (55)	212 (65)	300 (91)	390 (119)	510 (155)	750 (228)		

1 Excluding trap arm.
2 Except sinks, urinals and dishwashers.
3 Except six-unit traps or water closets.
4 Only four (4) water closets or six-unit traps allowed on any vertical pipe or stack; and not to exceed three (3) water closets or six-unit traps on any horizontal branch or drain.
5 Based on one-fourth (1/4) inch per foot (20.9 mm/m) slope. For one-eighth (1/8) inch per foot (10.4 mm/m) slope, multiply horizontal fixture units by a factor of 0.8.

Note: The diameter of an individual vent shall not be less than one and one-fourth (1-1/4) inches (31.8 mm) nor less than one-half (1/2) the diameter of the drain to which it is connected. Fixture unit load values for drainage and vent piping shall be computed from Tables 7-3 and 7-4. Not to exceed one-third (1/3) of the total permitted length of any vent may be installed in a horizontal position. When vents are increased one (1) pipe size for their entire length, the maximum length limitations specified in this table do not apply.

Table 10-2. (Continued)

Inch	mm
1-1/4	32
1-1/2	40
2	50
2-1/2	65
3	80

Plumbing Appliance, Appurtenance or Fixture	Min. Size Trap and Trap Arm[7]	Private	Public	Assembly[8]
Bathtub or Combination Bath/Shower	1-1/2"	2.0	2.0	
Bidet	1-1/4"	1.0		
Bidet	1-1/2"	2.0		
Clothes Washer, domestic, standpipe[5]	2"	3.0	3.0	3.0
Dental Unit, cuspidor	1-1/4"		1.0	1.0
Dishwasher, domestic, with independent drain	1-1/2"*[2]	2.0	2.0	2.0
Drinking Fountain or Watercooler (per head)	1-1/4"	0.5	0.5	1.0
Food-waste-grinder, commercial	2"		3.0	3.0
Floor Drain, emergency	2"		0.0	0.0
Floor Drain (for additional sizes see Section 702)	2"	2.0	2.0	2.0
Shower single head trap	2"	2.0	2.0	2.0
Multi-head, each additional	2"	1.0	1.0	1.0
Lavatory, single	1-1/4"	1.0	1.0	1.0
Lavatory in sets of two or three	1-1/2"	2.0	2.0	2.0
Washfountain	1-1/2"		2.0	2.0
Washfountain	2"		3.0	3.0
Mobile Home, trap	3"	12.0		
Receptor, indirect waste[1,3]	1-1/2"		See footnote 1,3	
Receptor, indirect waste[1,4]	2"		See footnote 1,4	
Receptor, indirect waste[1]	3"		See footnote 1	
Sinks				
Bar	1-1/2"	1.0		
Bar	1-1/2"*[2]		2.0	2.0
Clinical	3"		6.0	6.0
Commercial with food waste	1-1/2"*[2]		3.0	3.0
Special Purpose	1-1/2"	2.0	3.0	3.0
Special Purpose	2"	3.0	4.0	4.0
Special Purpose	3"		6.0	6.0
Kitchen, domestic	1-1/2"*[2]	2.0	2.0	
(with or without food-waste-grinder and/or dishwasher)				
Laundry	1-1/2"	2.0	2.0	2.0
(with or without discharge from a clothes washer)				
Service or Mop Basin	2"		3.0	3.0
Service or Mop Basin	3"		3.0	3.0
Service, flushing rim	3"		6.0	6.0
Wash, each set of faucets			2.0	2.0
Urinal, integral trap 1.0 GPF[2]	2"	2.0	2.0	5.0
Urinal, integral trap greater than 1.0 GPF	2"	2.0	2.0	6.0
Urinal, exposed trap	1-1/2"*[2]	2.0	2.0	5.0
Water Closet, 1.6 GPF Gravity Tank[6]	3"	3.0	4.0	6.0
Water Closet, 1.6 GPF Flushometer Tank[6]	3"	3.0	4.0	6.0
Water Closet, 1.6 GPF Flushometer Valve[6]	3"	3.0	4.0	6.0
Water Closet, greater than 1.6 GPF Gravity Tank[6]	3"	4.0	6.0	8.0
Water Closet, greater than 1.6 GPF Flushometer Valve[6]	3"	4.0	6.0	8.0

1. Indirect waste receptors shall be sized based on the total drainage capacity of the fixtures that drain therein to, in accordance with Table 7-4.
2. Provide a 2" (51 mm) minimum drain.
3. For refrigerators, coffee urns, water stations, and similar low demands.
4. For commercial sinks, dishwashers, and similar moderate or heavy demands.
5. Buildings having a clothes washing area with clothes washers in a battery of three (3) or more clothes washers shall be rated at six (6) fixture units each for purposes of sizing common horizontal and vertical drainage piping.
6. Water closets shall be computed as six (6) fixture units when determining septic tank sizes based on Appendix K of this Code.
7. Trap sizes shall not be increased to the point where the fixture discharge may be inadequate to maintain their self-scouring properties.
8. Assembly [Public Use (See Table 4-1)].

Air Gap

The plumbing code requires that certain appliances have an air gap to prevent water from siphoning out of a trap or from the appliance (see Figure 10-9). The air gap keeps the trap wet.

Cleanouts

A cleanout is a plugged tee installed in drain lines for allowing sewer pipes to be cleaned. When a line becomes plugged, the cleanout can be opened and a special coiled cable called a snake inserted into the pipe to break up whatever is causing the lines to be plugged. More information using snakes to clean out drain pipes is found in Chapter 17 on maintenance.

Backflow Preventers and Crossed Connection Control

Backflow preventers are installed in supply lines to prevent water from flowing backwards in them. This can happen in supply lines when the main valves are turned off. The problem with isolated lines is that contaminated water can sometimes be drawn into the fresh water supply lines.

A classic example of this occurs when using a hose in a swimming pool. If the water supply line is turned off and drained, the swimming pool water can be drawn back into the water supply lines. Then when the water supply line is turned back on again, the pool water, which may not be sanitary enough for drinking, is pushed on to the cold drinking water taps. Backflow preventers keep the lines charged and prevent water from being drawn back into the system.

Cross connection control is essentially a management activity. It means management of the system to prevent the wastewater lines and the fresh water lines from becoming interconnected. A cross connection control program is one where checks are made to verify the lines have not been crossed. There is an association of professionals dedicated to preventing crossed connection in water and sewer lines (see Chapter 19).

Chapter 11
Water Supply Systems

*N*ow that we understand the need for water purity and how supply and wastewater systems work, we can decide the best method of treating our facility's water supply. A variety of equipment is available to do the job—with many choices of systems that can improve our water's quality for occupant use.

WATER SYSTEM IMPROVEMENTS

To improve a facility's water supply, a number of equipment options are available. Water can be softened if it is hard, as well as filtered, distilled, cooled and so on. Each system is usually installed near where the water comes into a building or onto a complex. Some require an enclosure to protect the various pipes, vessels, valves, and instruments from the elements of harsh winters.

WATER SOFTENERS

Of all the water treatment systems, perhaps the water softener is the one that is most often encountered in facility management problems.

Hard Water
Calcium or magnesium will deposit on fittings and fixtures and on the insides of pipes. Water with these minerals in it makes it hard to get soap to form suds. Hence, the water is called hard water. These minerals leave a white scale residue when it dries or when water boils away in a pot. In addition to deposits, washing with hard water leads to stiff clothes and frizzled hair.

Depending upon the amount of hardness, which varies 6-120

149

grains per gallon, the economics of investing in a water softener can be paid back by savings in soap costs used in laundering and bathing.

As water hardness was explained in Chapter 4 here we will focus on equipment and methods used for water softening. The hardness is a dissolved salt-therefore, the method of softening is a chemical process where the undesired salt is exchanged with a salt that dissolves in water. In effect, the hard calcium carbonate is exchanged with sodium and becomes sodium bicarbonate; expressed as a chemical formula, this is:

$$CaCO3 \rightarrow 2NaHCO3.$$

Softeners

Water softeners (see Figure 11-1) are comprised of two or more vessels. The first vessel contains brine and the second vessel contains resin. The resin vessel contains beds of thousands of tiny beads called zeolite. The beads are sold in bulk form to the many water softener manufacturing companies throughout the world. Inside the water softener, the zeolite beads are coated with sodium chloride salts by rinsing the resin bed with brine made from sodium chloride salt solution. As soft water is drawn out of the water softener, the fresh hard water flows through the resin beds of zeolite where the sodium salts caked on the tiny beads exchange with the calcium salts in the water. The water coming out of the water softener is now "softened" because the calcium carbonate salts have been exchanged for sodium bicarbonate salts.

Periodically, the resin beads are regenerated with fresh sodium chloride brine, so a second tank called a brine tank is used to store brines before being drawn into the resin bed. During regeneration, the zeolite beads are flushed and rinsed with brine. The calcium, actually calcium chloride, is rinsed away in a drain cycle and the sodium salts in the brine recharge the resins for the next cycle.

People sensitive to the presence of sodium in their diet are sometimes cautioned to refrain from drinking water softened by this method. Hard water poses few health effects below 25 grains per gallon-however, in some areas of the country, the hardness can exceed 160 grains per gallon. When the level of dissolved salts in

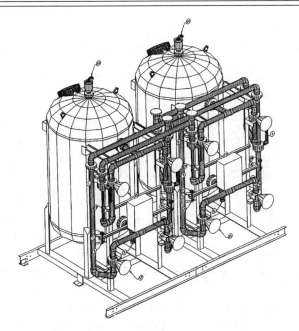

Figure 11-1. Typical water softener. Courtesy: Water and Power Technologies Systems, Inc., Salt Lake City, UT.

a water supply reaches these levels, it is necessary to remove them. Water softening will exchange sodium salts for calcium and magnesium salts, but the total volume of dissolved solids remains the same. To completely remove the salts, another process such as reverse osmosis is required.

Water softener manufacturers, using hardness sampling kits, can calculate the amount of "hardness" or the presence of calcium carbonates in the incoming water supplies. Then, by estimating the amount of water used by the facility per day, it is possible to calculate the number of cubic feet of resin that will be needed to soften that amount of water.

System Variables

Some of the variables that are faced by the water softener manufacturer are hardness of the incoming supply, volume used by the facility per day, peak flow of water at any one time in the day, and pressure of water. With these parameters specified by a facility manager, the facility can be confident the water softener meets the needs without being oversized.

In addition, the facility manager may want to inquire:

1. How much does the hardness vary over the year and has the equipment taken the variations into account?

2. How much does the pressure vary over the year, and has the equipment taken the variations into account? It is up to the facility to decide what water is to be softened. A survey sheet (see Figure 11-2) will aid the facility manager in overseeing the specification of water softeners for his facility.

In addition to the zeolite resin beds, the brine tank requires the periodic addition of salts to recharge the brines. There is no easy way to add salts. Most facility staffs haul the salt in to where the brine tank is on a fork lift or a pallet jack, and spend a few minutes or an hour or two cutting open the salt bags and dumping raw sodium chloride salts into the brine tank.

A number of maintenance problems occur when pieces of the paper bags fall into the brine tank and the facility staff does not remove it. The paper eventually plugs up the brine tank drain line and softening stops.

---

1. Is the water hard?

2. How hard is the water? _____ grains per gallon

3. What are the advantages of softening it that are desired?
    _____   Cost savings in soap reduction
    _____   Soft clothing and other washed items
    _____   Cosmetic enhancement of clean skin and hair

4. What size of a unit do I need based on flows and use?

5. Is there room for it in the present facility?

6. Does the room have the necessary drains, power and water supplies for this project?

Figure 11-2. Questions for determining the feasibility of water softening.

Options

Modern water softeners come with a variety of options designed to enhance the quality of water and reduce operating costs. Industrial water softeners have double resin beds. In this way, one resin bed supplies soft water to the building while the other one regenerates. Regeneration time depends upon the size of the bed but most are able to regenerate in less than three hours. Some facilities set timers on the water softeners to regenerate at night when water use in the facility is low.

Instrumentation

A water softener requires a control system to change the positions of valves for regeneration. The control system can work off a meter that measures the flow, a mechanical clock timer or electronic controls. As is always the case with electronics, either a battery backup should be supplied or the electronics will reset themselves after a power failure. This type of problem is common in many facilities and a facility manager can expect to find electronic controls malfunctioning on the water softener if he continues to receive a water problem call right after a power bump in his facility. Some softeners have only mechanical control that uses turbine meters and water pressure to control the cycles of the systems. These mechanical-only softeners do not require any electrical power with the inherent cost advantage of zero electric costs. However, most water softeners are located inside, and with the lights, it is a simple matter to provide power for the water softener control system.

Newer softeners use a hardness-monitoring probe inserted into the water pipe just downstream from the softener. The probe constantly reads the hardness and, when the resin bed is exhausted and needs regeneration, the probe signals the start of a regeneration Cycle. The probe can be sequenced with a clock that trips a switch that will not start the regeneration cycle until late the next night when the plant is not operating.

Maintenance Considerations

Facility management of water softeners consists of performing routine service on the equipment. Logs or records should indicate the staff is checking the softener daily and using a test kit to verify the readings of the hardness of both the incoming water and the softened water. The logs should also indicate when raw salt was added to the brine tank. Knowing the volume of water used and the volume from the brine tank for a regeneration cycle, it should be fairly easy to predict the time when salts are required to be added.

A facility manager should also be careful about scheduling maintenance of water softeners during certain high operating hours. The wrong settings and valve positions can lead to brine contamination in the building's water piping. This problem requires flushing of all the lines in the building by the staff and the embarrassment of having to tell the staff not to drink the water until the lines have been flushed. Fortunately, the brine, while it may have an unpleasant taste, is not a serious health risk to most individuals.

Water softeners loose about three percent of the resin bed per year of operation. Old water softeners can have the resins checked by a water softener vendor to confirm the adequacy of the resins.

Support Utilities

For facility managers contemplating the addition of a water softener system, planning needs to take into account whether the softener must be enclosed within a building, which is a requirement in cold weather climates as the lines, valves and beds will freeze during winter. Warm weather facilities can locate the softeners outside provided the brine tank is kept covered to prevent contamination by dust and wind blown debris. Outside installations should have lighting for maintenance and power outlets that are electrically ground-fault-protected since much of the maintenance around water softeners is "wet"-type work.

OSHA 1910 (Worker Safety Rules) will require confined space entry procedures be established for entering into tanks or vaults for maintenance. Chapter 17 provides a description of confined space entry procedures. Alternatives to sodium water softeners include use of other resins that exchange the calcium with non-sodium type materials or high pressure reverse osmosis units to filter out the calcium particles. Since reverse osmosis units do more than remove calcium ions, they are discussed separately later in this chapter.

FILTERS

Several types of filters are used to remove unwanted turbidity and particles. All filters use a media to capture the particles. Media types include sand, diatomaceous earth and fiber cartridge. Usually filtration is used in recirculating loops to remove unwanted items such as hair, sand, sticks, small stones and other particle matter.

Filters can be used as stand-alone items for particle removal only, or as a preliminary treatment step prior to more refined treatment. Filters are often used as a pretreatment step in chlorination or other water sterilization steps. In some utilities for small cities or towns, filtration and chlorination are all of the treatment provided for drinking water.

Sand Filters

In the rapid sand filter, sand and gravel are used as the filter medium. Figure 11-3 shows a rapid sand filter for a large swimming pool. Water flows down through the sand and gravel and the unwanted particles are captured on the sand particles. Periodically, the sand is backwashed, stirring up the bed and carrying away the materials that have been filtered out.

Figure 11-3. A rapid sand filter for a large swimming pool.

Rapid sand filters are limited to a fixed amount of flow per cubic foot of sand and a flow rate (speed) through the sand. Sand particle sizes are clearly defined in the beds of the filter. Usually, the fine sand is fixed on top of the large grained sand.

Like the water softeners, the sand and gravel beds are inside a pressurized tank.

Maintenance considerations for rapid sand filters are similar to considerations for softeners. Occasionally the sand bed has to be repacked with a necessary entry into the tank with bags of sand. A storm or sanitary drain is needed near the filter to allow the crews to drain the tanks. In cold climates, the filter should be enclosed and heating and lighting provided. Lastly, the electrical circuits should be protected with ground fault interrupter breakers to reduce the chances of electrical shock to maintenance personnel.

Diatomaceous Earth Filters

Similar to sand filters, a diatomaceous earth filter performs the same function and operates similarly. The advantage of diatomaceous earth is that overall size of the filter is usually smaller than a sand filter for the same flow of water. Diatomaceous earth is a very fine material, similar to clay or flour in consistency. Because of the fine particles, a greater flow per unit area or thickness is calculated compared with the sand. Diatomaceous earth filters are more difficult to operate than sand and are often used in areas of the country where sand is not abundant. Valves and backwashing is similar to sand filters. Figure 11-4 shows a diatomaceous earth filter for a small heated spa.

As with softeners and sand filters, maintenance and operations of diatomaceous earth filters require an enclosed facility in cold climates. However, as a result of the smaller size of the components for diatomaceous earth filters, the corresponding facility size can be smaller as well. Again, drains, heat and lighting are necessary and electrical outlets should be protected.

Both sand and diatomaceous earth filters require a large drain pipe for backwashing. The flow of the backwashing process should be known before installing the drain piping for the system—otherwise, backwashing can cause flooding of the room or equipment where it installed until the drain line can catch up with the backwash cycle.

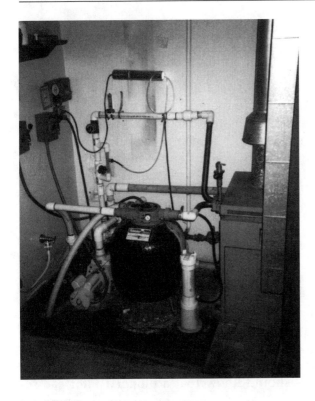

Figure 11-4. A diatomaceous earth filter used for a small heated spa.

Cartridge Filters

Small pools, tubs, and other systems use cartridge-type filters. The cartridge filter uses a fiber (usually paper) to filter out unwanted particles. The fiber cartridge is housed in a metal or plastic cylinder. Note that some local codes require that the filter housing cylinder be made from stainless steel. This has been thought to keep the water sterilized better than the plastic housing. The paper cartridge is chosen based upon the number of square feet per gallon per minute of flow.

Regarding maintenance, the paper cartridges can be removed and changed, but while the cartridge is changed, the water supply system has to either bypass the filter or the system has to be shut down. The cartridge can then be rinsed, washed down or replaced.

If the paper in the cartridge is torn, it will have to be replaced since unfiltered water will go through the hole at the tear. In terms of size, cartridge filters compare with diatomaceous earth filters.

As with the other filters, the facilities should provide for heat and light for maintenance during cold weather climates. Electrical equipment, if used, should be ground-fault-protected.

Filter Instruments

Most filters do not require the use of electronic instruments. Sand and Diatomaceous Earth Filters can be backwashed to flush out contaminants either manually by an operator or automatically on a clock timer. The clock and valve control mechanisms can be operated mechanically where electricity is not available, but since most modern installations are located near a power source, manufacturers have gone forward with electronic instruments for most installations.

Filter efficiency is measured with the use of a differential pressure gauge. The gage is set up to read the pressure in the water lines upstream and downstream of the filtration equipment. As the filter becomes plugged and fouled with debris to be filtered, the pressure needed to push the water increases.

Most filters are designed to operate with about a 1-2 pounds per square inch (psi) drop across the filter and to backwash or be cleaned when the pressure difference increases to 5-7 psi.

Facility managers should know that a plugged filter can severely lower downstream system pressures. If the downstream pressure gets too low, air can get into the piping creating flow and contamination problems. In addition, filters specified with a high pressure drop can lead to increased energy costs from pumping, making the filter a cost burden in facility operation.

CHLORINATORS

Facility managers do not get involved with chlorinators too often since most chlorination is performed by the utility. But on occasion, special chlorination is required. Chlorinators for facility managers are most often associated with pools, jacuzzi/hot tub baths or other public bathing facilities. When chlorinators are used in conjunction with filters, the filters work on the flows first, then the chlorinators add the chlorine.

A chlorinator will inject free chlorine into the water, with various benefits (see Chapter 4). Chlorination systems consist of bottles of chlorine gas, under pressure, that gently vent the gas into water supplies downstream from pump discharges. The chlorine gas reacts with free hydrogen radicals to make hydrochloric acid (HCL).

The free chlorine and the hydrochloric acid in the water kills living microorganisms and purifies the water supplies. Most chlorinators use a needle valve in the lines from the chlorine bottle to regulate the flow of the chlorine into the supply. The pump curve (see Chapter 8) is used to estimate the flow, and the needle valve injects an amount approximately equivalent to 0.8 milligrams per liter into the water. At this low level, the amount of hydrochloric acid is extremely small so that it is diluted by a large amount of water and the risks of too much chlorine are low.

However, the chlorine does form an acid and acid is corrosive. If the mixing is not done properly, therefore, too much or too little chlorine gets into the water. Too much, of course, will make the water acidic, while too little will not kill the microorganisms.

One of the reasons chlorine has been the standard for the past 100 years is that the window between where the chlorine is successfully purifying the water to where it is potentially harmful to the piping and to humans is relatively large compared to the use of other chemicals.

Other methods of chlorination used for small pools consist of the use of tablets containing chloramine. With this type, the chlorinator is a cylinder or tank filled with these tablets, and water flows through the tablets into the main water stream. This type of system reduces the amount of chlorine gas and it is easier to operate. However, the operator costs to fill the vessel with tablets and the cost of tablets compared to the cost of chlorine gas is expensive. Many swimming pools use tablets instead of free gas because the training level required for the operators is lower.

Chlorine Gas Management

The decision to use bottled chlorine gas should be based upon economics. Chlorine gas is a poison but many utilities use and handle it often and have experience with this chemical. Chlorine gas bottles should always be chained up to prevent tipping over and only the bottle "on line" should be open. The others should be locked closed. No more chlorine than is necessary should be on site at any time.

Finally, if free chlorine gas is used at the facility, it should be stored under lock and key and the operators should be trained in rescue techniques and to recognize a leak if one occurs. And the facility manager should make sure that he has *records* that training, quality control and tests (with results recorded) have been conducted. (See Chapter 3 for more information on OSHA rules and Chapter 17 for more information about maintenance.)

Chlorinator Maintenance

Facility managers should make sure that the staff has been trained in the safe operation and maintenance of chlorination equipment. Water tests should be taken and logged to provide a record of the adequacy of the water. In addition, if free chlorine is present at the facility, the managers should make sure that the Material Safety Data Sheets (MSDS) for chlorine are posted where they can be read by the staff.

Finally, because chlorine gas is poisonous, workers must be trained in safe rescue techniques and OSHA-approved Level A backpacks must be available for rescue. Facility managers should make sure that the staff has been *trained to use the rescue equipment*. It might not hurt to practice a drill periodically; it can be coordinated so that it does not interfere with business. Drills should not be conducted in front of clients or customers, unless of course, they want to see it. An MSDS for chlorine is shown in Figure 11-5. In addition, a simple contingency procedure has been provided for facility managers to see the simple, effective steps that can be taken in the event of a chlorine leak (see Figure 11-6).

Alternatives To Chlorinators

In the late 1980s, there was a movement in this country to stop the use of chlorine in water supplies because of some of these potential health risks. In addition, chlorine gas reacts with organic compounds in surface supplies to release trihalomethanes (THM) which are suspected carcinogens. However, alternatives to chlorine have still not proven as effective for similar costs.

Figure 11-5. Material Safety Data Sheet for Chlorine. Courtesy CHEMTREC.

MATERIAL SAFETY DATA SHEET   **ISSUED: 10/23/97**

CHLORINE   **REVISED: 11/01/99**

SECTION I - PRODUCT IDENTIFICATION

Westlake CA&O
2468 Industrial Parkway
P O Box 527
Calvert City, KY 42029

Telephone No.: (270) 395-4151
Transportation Emergency No.:
CHEMTREC: (800) 424-9300
Medical Emergency No.:
POISON CENTER: (216) 379-8562

Chemical Family: Halogen
Chemical Name/Synonyms: Chlorine
Trade Mark: None
Formula: Cl_2; (Cl-Cl)
C.A.S. Registry No.: 7782-50-5
TSCA Inventory Status: All ingredients are listed on the USEPA's TSCA inventory
Canadian Domestic Substances List Status: All ingredients have been nominated or are
eligible for inclusion.
Workplace Hazardous Materials Information System (WHMIS) Classification: C,E
Product Use: Various Applications
SARA 313 Information: This product contains a toxic chemical or chemicals subject to the
reporting requirements of section 313 of Title III of the Superfund
Amendments and Reauthorization Act of 1986 and 40 CFR part 372.

SECTION II - HAZARDOUS INGREDIENTS

Hazard Summary Statement: WARNING! HIGHLY TOXIC. CORROSIVE. May be fatal if
inhaled. Strong oxidizer. Most combustibles will burn in chlorine as they do in oxygen. Read
entire Material Safety Data Sheet (MSDS).

Material	C.A.S. Number	Amount in Product	ACGIH TLV-TWA	OSHA PEL-TWA
Chlorine [1,2,4,5,6]	7782-50-5	> 99.5%	0.5 ppm 1 ppm short term exposure limit (STEL)	1 ppm - ceiling

N.A. - Not Applicable   **N.E. - Not Established**

(*Continued*)

Figure 11-5. Material Safety Data Sheet for Chlorine (*Continued*). Courtesy CHEMTREC.

Legislative Footnotes

[1]Ingredient listed on SARA Section 313 List of Toxic Chemicals.

[2]Ingredient listed on the *Pennsylvania Hazardous Substances List.*

[3]Ingredient listed on the California listing of *Chemicals Known to the State to Cause Cancer or Reproductive Toxicity.*

[4]Ingredient listed on the *Massachusetts Substance List.*

[5]*Workplace Hazardous Materials Information System* ingredient found on the Ingredient Disclosure List - Canada.

[6]Ingredient listed on the *New Jersey Right to Know Hazardous Substance List.*

Notes:

TLV-TWA - Threshold Limit Value - Time Weighted Average guideline for concentration of the chemical substance in the ambient workplace air. (The skin notation calls attention to the skin as an additional significant route of absorption of the listed chemical.) American Conference of Governmental Industrial Hygienists (ACGIH).

OSHA PEL - OSHA Permissible Exposure Limit, 8-hour TWA. 29 CFR 1910.1000, Transitional Limits column, Table Z-1-A, Table Z-2, and Table Z-3.

SECTION III - PHYSICAL DATA

Appearance: Greenish-yellow gas
            or amber liquid
Odor: Pungent, suffocating bleach
            like odor
Percent Volatiles: >99.5
Solubility in Water: Slight
Physical State: Gas (liquid under pressure)

Specific Gravity: Dry Gas (2.48 @ 0°C)
                Liquid (1.47 @ 0/4°C)
Melting Point: -101°C (-150°F)
Molecular Weight: 70.9
Vapor Pressure: 73 psia @ 50°F
Vapor Density: 2.5 (Air=1)

SECTION IV - FIRE & EXPLOSION HAZARD DATA

Flash Point: Test is not applicable to gases. Not combustible. Chlorine can support combustion and is a serious fire risk.

Flammable Limits in Air: Not Applicable

Figure 11-5. Material Safety Data Sheet for Chlorine (*Continued*). Courtesy CHEMTREC.

Note:

Flash Point: The lowest initial temperature of air passing around the specimen at which sufficient combustible gas is evolved to be ignited by a small external pilot flame.

Extinguishing Media: For small fires use dry chemical or carbon dioxide. For large fires use water spray, fog or foam.

Special Firefighting Procedures: Wear full face positive pressure self-contained breathing apparatus (SCBA). Wear full protective gear to prevent all body contact (moisture or water and chlorine can form hydrochloric and hypochlorous acids which are corrosive). Personnel not having suitable protection must leave the area to prevent exposure to toxic gases from the fire. Use water to keep fire-exposed containers cool (if containers are not leaking). Use water spray to direct escaping gas away from workers if it is necessary to stop the flow of gas. In enclosed or poorly ventilated areas, wear SCBA during cleanup immediately after a fire as well as during the attack phase of firefighting operations.

Unusual Fire and Explosion Hazards: Chlorine and water can be very corrosive. Corrosion of metal containers can make leaks worse. Although non-flammable, chlorine is a strong oxidizer and will support the burning of most combustible materials. Flammable gases and vapors can form explosive mixtures with chlorine. Chlorine can react violently when in contact with many materials and generate heat with possible flammable or explosive vapors. Chlorine gas is heavier than air and will collect in low-lying areas.

Explosive Characteristics: Containers heated by fire can explode.

SECTION V - Reactivity

Stability: Stable

Hazardous Polymerization: Will not occur.

Hazardous Decomposition Products: Hydrogen chloride may form from chlorine in the presence of water vapor.

CAUTION! Oxidizer. Extremely reactive.

Incompatibility (Materials to Avoid): Chlorine is extremely reactive. Liquid or gaseous chlorine can react violently with many combustible materials and other chemicals, including water. Metal halides, carbon, finely divided metals and sulfides can accelerate the rate of chlorine reactions. Hydrocarbon gases, e.g., methane, acetylene, ethylene or ethane, can react explosively if initiated by sunlight or a catalyst. Liquid or solid hydrocarbons, e.g., natural or synthetic rubbers, naphtha, turpentine, gasoline, fuel gas, lubricating oils, greases or waxes, can react violently. Metals, e.g., finely powdered aluminum, brass, copper, manganese, tin, steel and iron, can react vigorously or explosively with chlorine. Nitrogen compounds, e.g., ammonia and other nitrogen compounds, can react with chlorine to form highly explosive nitrogen   trichloride.   Non-metals,

(MSDS - Chlorine)   Page 3 of 8

Figure 11-5. Material Safety Data Sheet for Chlorine (*Continued*). Courtesy CHEMTREC.

e.g., phosphorous, boron, activated carbon and silicon can ignite on contact with gaseous chlorine at room temperature. Certain concentrations of chlorine-hydrogen can explode by spark ignition. Chlorine is strongly corrosive to most metals in the presence of moisture. Copper may burn spontaneously. Chlorine reacts with most metals at high temperatures. Titanium will burn at ambient temperature in the presence of dry chlorine.

SECTION VI - HEALTH HAZARD DATA

Threshold Limit Value: See Section II.

Primary Routes of Exposure: Inhalation, skin and eye contact.

Effects of Overexposure:

Acute: Low concentrations of chlorine can cause itching and burning of the eyes, nose, throat and respiratory tract. At high concentrations chlorine is a respiratory poison. Irritant effects become severe and may be accompanied by tearing of the eyes, headache, coughing, choking, chest pain, shortness of breath, dizziness, nausea, vomiting, unconsciousness and death. Bronchitis and accumulation of fluid in the lungs (chemical pneumonia) may occur hours after exposure to high levels. Liquid as well as vapor contact can cause irritation, burns and blisters. Ingestion can cause nausea and severe burns of the mouth, esophagus and stomach.

Chronic: Prolonged or repeated overexposure may result in many or all of the effects reported for acute exposure (including pulmonary function effects).

Emergency and First Aid Procedures:

Inhalation (of process emissions): Take proper precautions to ensure rescuer safety before attempting rescue (wear appropriate protective equipment and utilize the "buddy system"). Remove source of chlorine or move victim to fresh air. If breathing has stopped, trained personnel should immediately begin artificial respiration or, if the heart has stopped, cardiopulmonary resuscitation (CPR). Avoid mouth-to-mouth contact. Oxygen may be beneficial if administered by a person trained in its use, preferably on a physician's advise. Obtain medical attention immediately.

Eye Contact: Immediately flush the contaminated eye(s) with lukewarm, gently flowing water for at least 20 minutes while the eyelid(s) are open. Take care not to rinse contaminated water into the non-affected eye. If irritation persists, obtain medical attention immediately.

Skin Contact: As quickly as possible, flush contaminated area with lukewarm, gently running water for at least 20 minutes. Under running water, remove contaminated clothing, shoes, and leather watchbands and belts. If irritation persists, obtain medical attention immediately. Completely decontaminate clothing, shoes and leather goods before re-use, or, discard.

Ingestion: Not an anticipated hazard.

Figure 11-5. Material Safety Data Sheet for Chlorine (*Continued*). Courtesy CHEMTREC.

SECTION VII - SPILL & LEAK PROCEDURE

Steps to be taken in case material is released or spilled: Restrict access to the area until completion of the cleanup. Issue a warning: POISON GAS. DO NOT TOUCH SPILLED LIQUID. Do no use water on a chlorine leak (corrosion of the container can occur, increasing the leak). Shut off leak if safe to do so. Wear NIOSH/MSHA-approved, self-contained, full-face, positive pressure respirator and full protective clothing capable of protection from both liquid and gas phases. Persons without suitable respiratory and body protection must leave the area.

The following evacuation guide was developed by the U.S. Department of Transportation (DOT): Spill or leak from a smaller container or small leak from a tank - isolate in all directions 250 feet. Large spill from a tank or from a number of containers - first, isolate 520 feet in all directions; secondly, evacuate in a downwind direction 1.3 miles wide and 2.0 miles long. Keep upwind from leak. Vapors are heavier than air and pockets of chlorine are likely to be trapped in low-lying areas. Use water spray on the chlorine vapor cloud to reduce vapors. Do not flush into public sewer or water systems. Chlorine can be neutralized with caustic soda or soda ash. Alkaline solutions for absorbing chlorine can be prepared as follows:

For 100 pound containers: 125 lbs. of caustic soda and 40 gallons of water
For 2,000 pound containers: 2,500 lbs. of caustic soda and 800 gallons of water
For 100 pound containers: 300 lbs. of soda ash and 100 gallons of water
For 2,000 pound containers: 6,000 lbs. of soda ash and 2,000 gallons of water

CAUTION: Observe appropriate safety precautions for handling alkaline chemicals. Heat will be generated during the neutralization process.

Waste Disposal Method: Due to its inherent properties, hazardous conditions may result if the material is managed improperly. It is recommended that any containerized waste chlorine be managed as hazardous waste in accordance with all applicable federal, state, and local health and environmental laws and regulations.

SECTION VIII - SPECIAL PROTECTION INFORMATION

Ventilation: Effective exhaust ventilation should always be provided to draw fumes or vapors away from workers to prevent routine inhalation. Ventilation should be adequate to maintain the ambient workplace atmosphere below the legislated levels listed in Section II.

Respiratory Protection: Use NIOSH approved acid gas cartridge or canister respirator for routine work purposes when concentrations are above the permissible exposure limits. Use full facepiece respirators when concentrations are irritating to the eyes. A cartridge-type escape respirator should be carried at all times when handling chlorine for escape only in case of a spill or leak. Re-enter area only with NIOSH approved, self-contained breathing apparatus with full facepiece. The respiratory use limitations made by NIOSH or the manufacturer must be observed. Respiratory protection programs must be in accordance with 29 CFR 1910.134.

Eye/Face Protection: Non-ventilated chemical safety goggles or a full face shield.

Figure 11-5. Material Safety Data Sheet for Chlorine (*Continued*). Courtesy CHEMTREC.

Skin Protection: Wear impervious gloves, coveralls, boots and/or other resistance protective clothing. Safety shower/eyewash fountain should be readily available in the work area. Some operations may require the use of an impervious full-body encapsulating suit and respiratory protection.

Note: Neoprene, polyvinyl chloride (PVC), Viton, and chlorinated polyethylene show good resistance to chlorine.

Additional: Do not eat, drink or smoke in work areas. Maintain good housekeeping.

SECTION IX - SPECIAL PRECAUTIONS

Material Handling: Do not use near welding operations, flames or hot surfaces. Move cylinders by hand truck or cart designed for that purpose. Do not lift cylinders by their caps. Do not handle cylinders with oily hands. Secure cylinders in place in an upright position at all times. Do not drop cylinders or permit them to strike each other. Leave valve cap on cylinder until cylinder is secured and ready for use. Close all valves when not in actual use. Insure valves on gas cylinders are fully opened when gas is used. Open and shut valves at least once a day while cylinder is in use to avoid valve "freezing". Use smallest possible amounts in designated areas with adequate ventilation. Have emergency equipment for fires, spills and leaks readily available. Wash thoroughly after handling product. Provide a safety shower/eyewash station in handling area. An emergency contingency program should be developed for facilities handling chlorine.

Storage: Store in steel pressure cylinders in a cool, dry area outdoors or in well-ventilated, detached or segregated areas of noncombustible construction. Keep out of direct sunlight and away from heat and ignition sources. Cylinder temperatures should never exceed 51°C (125°F). Isolate from incompatible materials. Store cylinders upright on a level floor secured in position and protected from physical damage. Use corrosion resistant lighting and ventilation systems in the storage area. Keep cylinder valve cover on. Label empty cylinders. Store full cylinders separately from empty cylinders. Avoid storing cylinders for more than six months. Comply with applicable regulations for the storage and handling of compressed gases.

SECTION X - HAZARD CODES

NFPA		HMIS	
(National Fire Protection Association)		(Hazardous Materials Identification System)	
Health:	4	Health:	3
Flammability:	0	Flammability:	0
Reactivity:	0	Reactivity:	0
Special:	OXY	Personal Protection:	X[*]

Key:
0 = Insignificant
1 = Slight   [*] See MSDS for specified protection
2 = Moderate
3 = High
4 = Extreme

(MSDS - Chlorine)   Page 6 of 8

Figure 11-5. Material Safety Data Sheet for Chlorine (*Continued*). Courtesy CHEMTREC.

USER'S RESPONSIBILITY

This bulletin cannot cover all possible situations which the user may experience during processing. Each aspect of the user's operation should be examined to determine if, or where, additional precautions may be necessary. All health and safety information contained within this bulletin should be provided to the user's employees or customers. Westlake CA&O Corporation must rely upon the user to utilize this information to develop appropriate work practice guidelines and employee instructional programs for his or her operation.

DISCLAIMER OF LIABILITY

As the conditions or methods of use are beyond our control, we do not not assume any responsibility and expressly disclaim any liability for any use of this material. Information contained herein is believed to be true and accurate but all statements or suggestions are made without warranty, expressed or implied, regarding the accuracy of the information, the hazards connected with the use of the material or the results to be obtained from the use thereof. Compliance with all applicable federal, state and local laws and regulations remains the responsibility of the user.

Figure 11-5. Material Safety Data Sheet for Chlorine (*Continued*). Courtesy CHEMTREC.

SHIPPING INFORMATION

IDENTIFICATION - DOMESTIC TRANSPORTATION

Proper Shipping Name (172.101(c)): **Chlorine**
(Technical Name(s)) 172.203(k): **N/A**
Hazard Class 172.101(d): **2.3**
UN/NA# 172.101(e): **UN 1017**
Haz. Substance 171.8: **RQ (Chlorine)**
Reportable Quantity (Appendix A to 172.101): **10 LB**
Inhalation Hazard 172.2a(b): **Zone B, Poison-Inhalation Hazard, Marine Pollutant**
Package Code 172.101(f): **N/A**
Placarded: **Poison Gas**

PACKAGING (Part 173)

♦ Packaging Section (172.101(i)) - Col. 8(a):   None
                                    Col. 8(b):   173.304
                                    Col. 8(c):   173.314, 173.315

♦ General Packaging Section - General 173.24   Hazard Class: **POISON GAS**

MARKING

A. Proper Shipping Name (172.301(a))   (Technical Name) (172.301(b))
B. UN/NA Number (172.301(a))
C. Name & Address (172.301(d))
D. THIS END UP (172.312(a))
E. Hazardous Substance RQ (Name) (172.324)
      ORM Designation (172.316(a))
      Inhalation Hazard (172.313(a))

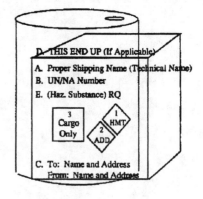

DOMESTIC LABELING
1. HMT LABELS (172.400)
2. Additional Subsidiary Hazard (172.402(a)):
                          8 (Corrosive)

DANGEROUS GOODS DETERMINATION (38th Edition) IATA

♦ Air Transport of This Material if Forbidden (Passenger and Cargo)

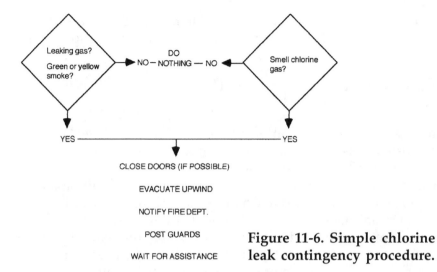

Figure 11-6. Simple chlorine leak contingency procedure.

Several other types of water sterilizers are available that will be briefly mentioned here, including ozone and UV light purifiers.

Ozone. Ozone, like chlorine, can be injected into water supplies and performs the same functions. And like chlorine, ozone will make the water acidic, thereby killing the living microorganisms. Ozone is not as poisonous as chlorine, but still requires safe handling procedures similar to chlorine.

Maintenance of ozone systems requires staff training and records. Since both ozone and chlorine are reactive gases, the rooms where they are used should be well-ventilated, and personal protective equipment—including safety goggles, gloves and other equipment—must be made available for the facility staff.

UV Light Purifiers. As an alternative to chemical treatment to purify water with gas, ultraviolet (UV) light can be used. Water is passed through a clear plastic tube where ultraviolet bulbs have been placed shining the UV light through the clear wall and into the water. The UV light kills microorganisms. Unfortunately, UV light requires electrical power, presenting another operating cost component, while the other systems do not. Other drawbacks to UV light water purifiers include turbidity. If the fluid is not clear, the UV light will not pass all the way through and sterilize the whole stream. Another drawback is that the UV light does not work beyond the clear plastic section. Note that another advantage of chlorine and ozone injection over UV light is that there is some residual treatment that carries over into the downstream piping.

REVERSE OSMOSIS UNITS

Similar in size, shape and style to the cartridge filter unit, a reverse osmosis unit can also be used to purify water. Reverse osmosis works differently from chemical or ultraviolet protection.

The heart of a reverse osmosis (RO) unit is a cartridge/membrane that has very fine pores. The RO unit is sized to pass pure water and to prevent other molecules such as salts, carbonates, microorganisms and similar big objects from passing through the membrane. If there is not enough water pressure naturally in the system to force the water through the reverse osmosis membrane, a powerful pump is required to give the water enough energy to go through the reverse osmosis unit.

Problems associated with RO units include the high energy costs to force the water through the membrane and the waste water stream necessary to strip the membranes. Because RO units are, in effect, an extremely fine filter, a lot of water is lost when backwashing the membrane to remove the filtered materials. Lastly, the membranes can sometimes become ineffective. Water operators call this "poisoning" the membrane. That is, the membranes pores become clogged with material and cannot be cleared, causing increased energy costs. Sometime the membrane itself becomes contaminated with microorganisms and adds, rather than removes, live bacteria.

RO units will remove water hardness, rather than letting it be replaced with sodium, so for occupants concerned about low sodium diets, an RO unit will soften the water as well. Usually to protect the RO unit membrane, one of the other filter types is installed upstream. Some living microorganisms can get through the RO unit filter. Most notably, this includes viruses.

DEIONIZERS

Another method of treatment of specialty water is a deionizer sometimes called a demineralizer.

A deionizer looks similar to a twin-bed water softener, and it works on much the same principles. Like a water softener, a deionizer uses resins through which water flows and the resins exchange or remove the ions from the stream. When the ions in the supply water are removed the water has been deionized, hence the name. There are two basic types of deionizers, named for the types of resins used.

The first type is usually two vessels with separate types of resins. The two types are cation and anion resins. The first resin vessel removes the positively charged ions (the anions) and the second removes the negatively charged ions (the cations.)

The other type of deionizer is a mixed-bed deionizer that has both the cation and anion resins mixed in one vessel. The mixed-bed unit is smaller and is even used to polish water from the single-bed deionizer. As with all deionizers, the water should be run through a reverse osmosis unit before being sent to the deionizer to protect the life of the resin bed.

Deionized water is used in laboratories and in making sterile medical devices. It is also used in manufacturing computer/silicon/wafer chips. This is because the presence of ions in the rinse water can ruin the results of months of laboratory work or years of silicon chip manufacturing.

Since deionized water has had all of the ions removed, the water becomes a very poor conductor of electricity. This inability to act as a conductor, when water is usually a good conductor, is the method by which deionizers are specified and by which operations personnel can verify the unit is working properly.

As with all previous discussions on instrumentation and operating records, data should be collected on a routine basis. The records should include instrument readings from the unit. And the instruments on the unit should be checked regularly to verify they are reading properly.

The test for deionization is performed by measuring the resistance to electricity passing through the water. The industry has even gone so far as to coin a phrase for deionized water quality readings. The micromho (pronounced "mike-row-mow") is the inverse of reading of the ohm measuring the resistance. Good deionized water is any water reading better than 15 micromhos.

While regeneration of water softeners is performed with

MSDS and The Worker's Right To Know Law

The Worker's Right To Know Law (OSHA 1910) provides that employers and managers have the obligation to make sure that workers know what chemicals are used and what the dangers to workers potentially may be. In addition, workers are to be provided with access to this information and necessary personal protective equipment must be provided to the workers around chemicals that are potentially hazardous.

Reading MSDS and for deciding what personal protective equipment is necessary, is usually the role of the safety department but facility managers should know that the real professional in this area is the certified industrial hygienist. Industrial hygienists are trained in exposure limits, time-weighted averages, personal protective equipment and other worker safety-related interpretations.

The MSDS, provided usually by the manufacturer, provides a method of determining the hazards and risks of the chemicals. The MSDS lists the names of the chemical, the hazards of the chemical (whether corrosive, toxic, flammable or physical), the exposure levels and the recommended personal protective equipment.

Unfortunately, MSDS have become a type of legal document, protecting the manufacturer from all types of liability. As such, they are becoming increasingly difficult to read and trying to grab an MSDS in a contingency drill or real emergency and read and respond in a timely fashion is difficult at best. The MSDS is best used for training and reference. Most plants prepare several thick 3-ring notebooks that contain all of the MSDS for the site and post them in conspicuous places throughout the plant where workers con access and refer to the specific chemical being used. It is not recommended the book be posted where the public can access it, however, since portions of it may disappear over time (unless posted in an inaccessible place, such as a glass-covered cabinet holding a bulletin board). Supervisors and managers should make sure the MSDS are posted and be familiar with the contents.

brines, regeneration of twin-bed and mixed-bed deionizers is accomplished using strong acids and bases. For the base, the regeneration fluid is usually 18 percent sodium hydroxide (NaOH), which has a pH of 14, while for the acid resins, regeneration liquid can either be hydrochloric (HCL) or sulfuric acid (H2SO4). The acids have a pH of 1. Needless to say, these strong acids and bases are hazardous chemicals and workers servicing these units need to be carefully trained to handle the chemicals and to respond to the spills should they occur.

The facility manager should be careful to make sure the area around the base of a demineralizer/deionizer is diked to prevent any spills of strong acids or bases from washing into the sanitary sewer. The strong chemicals will kill the microbes at the sewage treatment plant, and there are severe penalties for allowing this kind of hazardous waste to be released into the environment.

Maintenance Considerations

The physical size of a deionizer or demineralizer will vary with the flows required. Depending upon the application, a demineralizer could be larger than the water softener for the same flow. In general, a demineralizer will be fairly small and could even be small enough fit on a desktop or similar-sized area. As with most of the other water quality enhancers and treatment units, a deionizer should be indoors to protect the equipment from inclement weather effects and to allow maintenance personnel to work on the units if necessary during periods of winter weather and snow and ice conditions. Drains for washdown should be provided—although for a deionizer, the drain should be protected just in case there is an acid or a base spill. Lighting and electrical power should be available on ground-fault-protected circuits. Operations logs and records should be maintained and recorded.

STILLS

A still remains, today, one of the most effective water purifiers for removing both organic, inorganics and microorganisms. In a still, water is heated until it vaporizes, and the resulting steam is collected on the walls of a condensing vessel or in collection coils downstream, where it is cooled and condenses back into water.

Under this method of treatment, the undesirable components are either left behind or cook off early and do not condense back into the water because of a different volatility. Organics such as pesticides remain a gas and require cooling to a lower temperature than water so the materials that cook off do not condense back into the stock.

The disadvantages of a still are its high energy costs associated with vaporizing the water either with electrical energy or by burning hydrocarbons (coal, oil, gas, wood). These high energy costs limit the use of stills to very specified uses. Some laboratories utilize a still for making pure, reagent grade water for research.

Stills have to be dismantled and cleaned periodically but maintenance and repair costs with still-type units are relatively low. More maintenance time is spent on the burners, fuel, or electrical supply than on cleaning the still.

The hazard of operating a still is with the heat since steam and the materials associated with a still get very hot and the possibility of injury exists from maintenance personnel working on the still before it has completely cooled and getting burned.

Use of homemade stills and other non-engineered equipment has also resulted in injury when the pressure from steam that results from boiling water exceeds the strength of the vessel. The vessel ruptures and the hot water spills and sprays violently. As a protective device stills have temperature and pressure controls similar to hot water systems.

Personal protective equipment when working around a still includes leather gloves, goggles, aprons and a face shield.

Instrumentation for stills includes temperature gauges and pressure gauges since the manager needs to know if boiling rates and condensation rates are being maintained. Elevation is also an important consideration in still operation since water boils and condenses at different temperatures depending upon the elevation. Cooling is usually accomplished with water being fed into the still—this way, the incoming water is being preheated by the steam gases before flowing into the still to be vaporized.

Water Coolers

Principally used for drinking, a facility will install a water cooler, usually near the restroom because the plumbing is close to the outside wall at that point. Water coolers are usually wall-mounted and are plumbed in with a supply of cold water. The cooler uses a refrigeration unit to cool the water as it comes in and distributes it up through a fountain. A drain line is provided from the water cooler to the sewer system. Little effort is required to maintain a water cooler but they should be routinely checked to make sure their electrical connections are sound. Ice makers function in a similar way to water coolers and require the same utilities.

Chapter 12
Hot Water Systems

*E*very facility manager wants to provide hot water effectively and at low cost. In most facilities, the absence of hot water for a short period, while inconvenient, is not disastrous. Hospitals are, of course, another matter since the hot water is necessary for patient bathing and controlling sanitary conditions. Proper design, planning, and a good understanding of the system provides hot water that meets the facility needs economically.

INTRODUCTION

For existing facilities, the hot water system, if it has been designed correctly, will be adequate. In an older facility, the manager may be concerned about the status of the system. This concern may stem from observations of water and rust-colored stains on the floor and piping in the hot water tank room. Another area of concern may be customer calls about the temperature of the water.

The callers usually say the water is not warm enough, but there may be the occasional call that the water is too hot. Perhaps a simple thermostat adjustment will solve the problem for now. Unfortunately, in today's busy and often hectic world of facility management, hot water system issues are given low priority. Once the facility manager obtains a little breathing room on other issues he can look into hot water supply problems or, if he can obtain the necessary funds, he can contact an engineering firm to perform a short study to address hot water needs. Figure 12-1 lists a series of quick assessment questions that allow a manager to determine the status of his hot water system. Armed with this information, the facility manager can decide whether to expend his resources on his hot water system immediately or wait a few weeks or months if the situation is sufficiently under control.

177

Figure 12-1. Hot water system assessment.

1. How old is the water heating system in the facility   _____ Years

2. Any complaints from occupants?

    _____ Too hot   _____ No. of calls

    _____ Too cold   _____ No. of calls

3. An inspection of the water heater tanks and equipment reveals... (check all that apply.)

    _____ Loose, fallen tank insulation

    _____ Red stains on pipe, insulation or floor

    _____ Wet and humid conditions in the room

    _____ Musty odors or conditions

    _____ Dry conditions

4. Are the gauges and thermometers working properly?

    _____ Yes   _____ No

    Pressure   _____

    Temperature   _____

5. Are there any records or logbooks for the system?

    _____ Yes   _____ No

    Last preventive maintenance inspection?   _____

    Last check of the fuel burner or system by utility?   _____

6. Any previous studies?

    _____ Yes   _____ No

    If "yes," was there any action taken, and what were the results?

    _____

    _____

    _____

To adequately meet facility needs, the hot water system must be sized correctly and the water should be the right temperatures to provide enough hot water for the needs of the users.

The system uses energy to heat the water, either a fuel or electricity, and the tanks and pipes will be insulated to conserve energy once the water has been heated. The system must meet certain codes and standards for the safety of the building occupants and for the operators and maintenance people who work on the system.

OPTIONS AVAILABLE TO THE FACILITY

Does the system supply enough of the right temperature of water to meet the facility needs? That is the question.

Hot water system design is a function of four basic elements that are interrelated, including 1) size of the tank, 2) its volume or dimensions, 3) flow of hot water in the pipes to the point where it is used and 4) the temperature of the heated water. Another variable that must be addressed by a facility manager is the energy source used for heating the water.

Most of the systems in the United States are hot water *storage* systems. In a domestic hot water storage system, a sufficient volume of water is heated in a tank where it waits until needed.

Turning on the hot water tap, or starting the washer or dishwasher, are typical uses for hot water. In a storage system, hot water flows out of the tank until either the user has enough hot water and valves are closed, or the water heater runs out of hot water.

As a facility grows, more and more users draw more hot water from the tank until finally there is not enough hot water to go around. The first thing the maintenance staff does is raise the tank's thermostat. Now the hot water in the tank is hotter so more users can draw bath and shower water because they will turn down the hot water faucet a little. However, if the water is too hot, careless users can be scalded.

The facility manager's objective is to meet the needs of the facility users as effectively as possible. Effective means cost- and energy efficient. The facility could have a large tank that heats water slowly, or a small tank that heats water quickly. The trade-offs are that with a large tank heating slowly, there is plenty of hot water, but the tank and the room for the tank are expensive. If the water is not used, energy is wasted while keeping the water warm while waiting for demands.

The other problem with a tank system is keeping the water warm in a long run of pipe between the user and the tank. The user turns on the tap and hot water does not come out for several seconds while the water between the tank and the tap slowly recharges with hot water again. Here, hot water is wasted when it sits idle and cools. In addition, allowing the water to run wastes it as well.

At the other extreme, a device called a "tankless" heater is put at the end of the pipe. With this heater, the water is not heated until it is ready to be used—but now, the fuel to heat the water has to be run to the heater. If the facility is large and spread out, multiple fuel lines and heaters have to be placed throughout the complex.

The decisions for hot water system trade-offs requires professional judgment. The judgment is made initially when the facility is designed by somebody who uses codes and standards to determine which options will work best for the facility. In most areas of the United States, designers use the *ASHRAE Handbook*, which is published by the more than 100-year-old American Society of Heating, Refrigerating and Air-Conditioning Engineers of Atlanta, GA.

Data from research and other published sources are used to determine how much hot water is needed for the items that require hot water use. A dishwasher, a shower, a hot meal, etc. are all typical examples of where hot water is used. Table 12-1 provides a guide for how much hot water is needed in various types of facilities, and Table 12-2 provides an estimate of domestic use for various hot water systems like dishwashers. Since the temperature for use varies with device, typical temperatures are also recommended by ASHRAE. The temperatures presented in Table 12-3 provide the facility manager with guidance for design temperatures.

Another area of trade-offs that require judgment is in the choice of fuels for heating the water. As with the choice of large or small storage, decisions are made whether fuel or electricity will be used for heat. If fuel is used, which fuel? Natural gas, LPG and fuel oil are common. Finally, solar energy can also be used.

Choice of fuels is an energy question. Facility managers should be aware that hot water systems use little fuel relative to other energy uses in a facility. As little as 6-7 percent of the facility manager's budget for energy consumption could be expended for hot water heating for domestic use.

Finally, the facility is subject to some public code requirements if direct flame is used for heating the water. The codes require adequate amounts of fresh air for the flame and for a path of exhaust for the products of combustion in order to protect the building's occupants from carbon monoxide poisoning and fire.

The facility manager must keep these factors in mind while determining the best way to upgrade a hot water system. The facility manager must decide which resources to utilize in conducting the evaluation.

SYSTEM SIZING AND DESIGN

Storage, variations of storage and recirculation and instantaneous supply are the three main methods of providing hot water. Each is discussed in the following paragraphs in order to inform the facility manager of the factors that must be considered in the system design.

Storage

The advantage of storing domestic hot water is that it provides an excellent combination of even temperatures and sufficient volume to respond to sudden peak demands.

With a storage system, the water is heated over a steady period of time and as demands are made on the supply, cold water flows into the tank to replace it. A general rule of thumb for tank sizing is that of the total tank capacity, only about 70 percent is usable before the incoming cold water drops the temperature significantly. For a storage system, ASHRAE provides guidance for sizing hot water tanks in typical facilities.

Table 12-1. Probable hot water demand for various facilities. Source: American Society of Heating, Refrigerating and Air-Conditioning Engineers (ASHRAE), *ASHRAE Handbook*, 1991.

On an hourly and daily, basis, normal use

Type of building	Maximum hour	Maximum day	Average day
Men's dormitories	3.8 gal (14.4 L)/student	22.0 gal (83.4 L)/student	13.1 gal (49.7 L)/student
Women's dormitories	5.0 gal (19 L)/student	26.5 gal (100.4 L)/student	12.3 gal (46.6 L)/student
Office buildings	0.4 gal (1.5 L)/person	2.0 gal (7.6 L)/person	1.0 gal (3.8 L)/person
Food service establishments:			
Type A-full-meal restaurants and meals/day* cafeterias	1.5 gal (5.7 L)/max. meals/h	11.0 gal (41.7 L)/max. meals/h	2.4 gal (9.1 L)/avg.
Type B-drive-ins, grilles, luncheonettes, meals/day* sandwich and snack shops	0.7 gal (2.6 L)/max. meals/h	6.0 gal (22.7 L)/max. meals/h	0.7 gal (2.6 L)/Avg.
Apartment houses: no. of apartments			
20 or less	12.0 gal (45.5 L)/apt.	80.0 gal (303.2 L)/apt.	42.0 gal (159.2 L)/apt.
50	10.0 gal (37.9 L)/apt.	73.0 gal (276.7 L)/apt.	40.0 gal (151.6 L)/apt.
75	8.5 gal (32.2 L)/apt.	66.0 gal (250 L)/apt.	38.0 gal (144 L)/apt.
100	7.0 gal (26.5 L)/apt.	60.0 gal (227.4 L)/apt.	37.0 gal (140.2 L)/apt.
200 or more	5.0 gal (19 L)/apt.	50.0 gal (195 L)/apt.	35.0 gal (132.7 L)/apt.
Elementary schools	0.6 gal (2.3 L)/student	1.5 gal (5.7 L)/student	0.6 gal (2.3 L)/student*
Junior and senior high schools	1.0 gal (3.8 L)/student	3.6 gal (13.6 L)/student	1.8 gal (6.8 L)/student*

*Per day of operation.

Table 12-2. Approximate requirements of hot water for various fixtures and devices. Source: American Society of Heating, Refrigerating and Air-Conditioning Engineers (ASHRAE), *ASHRAE Handbook*, 1991.

Demand is given in terms of gallons (liters) per hour per fixture, calculated at a final temperature of 140°F (60°C).

Fixture	Apartment house	Hotel	Office building	Private residence	School
Basins, private lavatory	2(7.6)	2(7.6)	2(7.6)	2(7.6)	2(7.6)
Basins, public lavatory	4(15)	8 (30)	6(23)		15(57)
Bathtubs	20(76)	20(76)		20(76)	
Dishwashers*	15 (57)	50-200		15(57)	20(100)
Kitchen sink	10(38)	30(114)	20(76)	10(38)	20(76)
Laundry, stationary tubs	20(76)	28(106)		20 (76)	
Pantry sink	5 (19)	10(38)	10(38)	5 (19)	10 (38)
Showers	30(114)	75(284)	30(114)	30(114)	225(850)
Service sink	20(76)	30(114)	20(176)	15 (57)	20(76)
Circular wash sinks		20(76)	20(76)		30(114)
Semicircular wash sinks		10(38)	10(38)		15 (57)
Demand factor	0.30	0.25	0.30	0.30	0.40
Storage capacity factor†	1.25	0.80	2.00	0.70	1.00

*Dishwasher requirements should be taken from this table or from manufacturer's data for the model to be used, if this is known.

†Ratio (of storage tank capacity to probable maximum demand per hour. Storage capacity may be reduced where an unlimited supply of steam is available.

Table 12-3. Representative hot water temperatures. Source: American Society of Heating, Refrigerating and Air-Conditioning Engineers (ASHRAE), ASHRAE Handbook, 1991.

	Minimum temperature	
Use	°F	°C
Lavatory		
Hand washing	105	40
Shaving	115	45
Showers and tubs	110	43
Commercial and institutional laundry	180	82
Residential dishwashing and laundry	140	60
Commercial spray-type dishwashing as required by National Sanitation Foundation		
Single or multiple tank hood or rack type		
Wash	150 min	65 min
Final rinse	180 to 195	82 to 90
Single-tank conveyor type		
Wash	160 min	71 min
Final rinse	180 to 195	82 to 90
Single-tank rack or door type		
Single-temperature wash and rinse	165 min	74 min
Chemical sanitizing glasswasher		
Wash	140	60
Rinse	75 min	24 min

The facility manager typically delegates the authority to the designer to decide which sets of curves and standards to recommend. A typical design for a hot water system would be a bath/change house in a factory.

Recirculation

To eliminate waiting while the hot water recharges the lines between the storage tank and the tap, large facilities install a recirculation loop from the end of the supply line back to the hot water storage tank. This keeps the lines between the tank and the tap charged with hot water at all times. The disadvantage to this option is additional costs to run the return pipe back to the tank.

In many systems, this scenario requires an additional pump to push the hot water around the loop. The pump, of course, has to have electricity, which requires additional wiring. A recirculation hot water system is shown Figure 12-2.

Instantaneous "Tankless" Heaters

The facility manager has the option to select an alternative method other than storage. The most commonly presented alternative is called instantaneous point-of-use heating. Here, the heat needed to increase the water temperature is added to the water immediately before it reaches the tap. Under this option, the hot water is available on demand when called for at the tap and little energy is wasted in heating water supplies that are not used. Disadvantages to this type of system are in providing an energy source to heat the water at the point of use. Either electricity or fuel must be supplied to wherever the heater is located. If fuel is supplied, fresh air and an exhaust vent must be provided as well. If electricity is the source, additional wires and outlets may be needed if the unit requires more than the traditional electrical outlet.

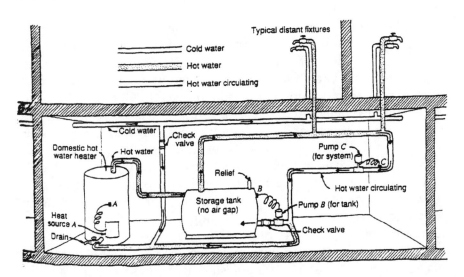

Figure 12-2. Recirculation hot water system. Courtesy: Stein and Reynolds, *Mechanical and Electrical Equipment for Buildings,* **8th Edition. ©1992, John Wiley & Sons, Inc.**

To get the energy into the water quickly, some of these tankless heaters divide the flow into very fine streams, akin to the radiator of an automobile. By dividing the flow to make it easy to get the heat into the water, energy is required to push the water through the fine tube fins—so much so that installation of a tankless heater can restrict the total flow to the user.

HEAT

When the facility manager knows how much hot water is needed and where it should be provided, the facility is ready to consider how to best provide the heat (and energy) to the hot water system.

Anywhere there is heat in a facility, hot water can be produced. If the plant is making steam or hot water for building heat, a portion of this heat can be used to heat domestic hot water if there is enough excess capacity in the heating equipment.

Heat, in the form of electricity, can be brought to the hot water tank or to the tankless heaters, provided there is enough power in the electrical system.

Fuel can be provided to where the water is going to be heated. Fuel can be anything that burns. See Table 12-4 for a list of the heating values in most common fuels. For most facilities the fuel selected will be a fuel commonly used in that area. Either fuel oils, natural gas, or liquefied petroleum gas (LPG) can be brought to where the water is to be heated. If fuels are to be burned inside a building, then fire and ventilation codes apply to the rooms or portion of the building where the heating equipment is installed. Normally, the costs of fuels for heating are much less than electrical energy costs, and it is economical to provide the necessary clearances, ventilation, fire protection and air quality permits in order to burn fuels instead of heating with electricity—but this decision is subject to economic evaluation (see Chapter 15).

Finally, energy from the sun can be used to provide hot water as in a solar hot water heating system if there is enough space at the facility for placing the solar collectors and if there is enough direct sunlight.

One of the major advantages of electrical or fuel-type supplies is the heat is available *on demand*. In the case of heat recovery or solar energy, the heat is available when the heat is on or when the sun is shining, respectively. Many facilities who heat domestic water with the same boilers that provide winter heat find they do not have hot water in the summer when the heating boilers have to be shut down for maintenance. Most facilities get around this problem by providing backup boilers. When one is running, the other is idle. If the first breaks down, the second one starts up to provide the necessary heat.

PIPING

After the water is heated, there remains the problem of getting it to where the users want it. The same principles for plumbing cold water apply to hot water.

Friction losses, the loss of pressure with distance and water hammer are exactly the same for hot and cold water systems. Hot water must have enough pressure to arrive at the tap in sufficient flow to meet the user needs. If the flow is restricted, the temperature of the water can be raised, giving the same results—but again, if the hot water alone is too hot, users can scald themselves. The facility manager has a responsibility to keep the temperatures under control.

After the correct volume of the right temperature water has been generated, it still has to be delivered to the tap at the right pressures and flows. For small facilities and for domestic use, the incoming cold water provides enough natural pressure to force the hot water from the tank, through the pipes and to the users. However, there is often less pressure in the hot water system, the result of losses in the pipes and tanks.

For a simple experiment, turn the cold water on full at a residence. Now turn the cold water off and turn the hot water on full. The slower release of hot water is the result of pressure lost in the hot water supply.

Pumps are added to the hot water system to make the pressures more equal. When taps are turned on, the pumps sense reduced pressure in the tank and boost the hot water supply.

Mostly for ease of installation hot water is piped with the

Table 12-4. How energy of fuels is figured to heat water.

Energy for heating water in the English system of units is measured in British thermal units (Btu). A Btu is the amount of energy required to heat one pound of water one degree Fahrenheit. Therefore, if one pound of water is 40 degrees and we want to heat it to 140 degrees, it would take 100 Btus to do it. Fuels are rated by the number of Btus that are released when heated. So when a fuel burns, the heat released is known.

The science of studying fuels and heat released is called combustion engineering. It requires a lot of chemistry and math to work out the details because it boils down to figuring out how much heat is released at the molecular level.

Fuel	Heat Content
Propane	90,000 Btu per gallon
Natural Gas	1,000 Btu per cubic ft.
Wood	8,800 Btu per pound dry, 7,000 Btu/lb. at 20% moisture content
Fuel Oil	19,000 Btu per pound, or 142,500 Btu per gallon
Coal	10,000 Btu per pound (average)
Kerosene	19,810 Btu per pound, or 135,000 Btu per gallon

same type of pipe and the same diameter as the cold water. This is a practical solution since engineering calculations will yield a larger pipe size for the hot water to make up for the additional losses. However, when installing the pipe, if one is a different size from the other, the plumber's efforts increase significantly since he must plumb different pipes for the two lines.

Hot water pipe is usually copper because it has been shown to be cost effective, easy to install, and most generally meets the system needs. Copper hot water pipe is sufficient up to and including the 3-inch size. Above this diameter, there are problems with fitting the joints together because typical field torch used to heat the joint is not big enough to provide uniform temperature throughout the entire fitting. For hot water piping above 3 inches in diameter, steel pipe is usually used. However, hot water piping that requires pipe in excess of 3-inch diameter would be rare and re-evaluation would be recommended.

Because hot water piping temperatures range between 105°F and 180°F, other types of pipes are not generally recommended. Plastic pipe can meet the requirement, but plastic tends to soften when warmed unless special resins are used which make the plastic hot water pipe more expensive than copper pipe.

Galvanized pipe (pipe coated with zinc) is not recommended for hot water systems because a chemical reaction takes place between the zinc and water at elevated temperatures.

Hot water pumps have internal workings resistant to hot water and are usually bronze or brass.

INSULATION

Once the water is hot, it should immediately be used. Thermal insulation, however, can be provided to keep the water warm and reduce the loss of heat. Several types of insulation are used, with foam and fiberglass being the two most common types. Usually, the insulating material is coated with a protective outer layer to protect the insulation from damage.

The thickness of the insulation is a function of the material's ability to resist heat transmission away from the material in contact with the heated water.

Most insulation materials are essentially air-encapsulating. That is, air is trapped in the insulation material and it is the air that acts as the insulating medium.

In the past asbestos was used as an insulating material on hot water tanks and pipes. Many older facilities had significant amounts of asbestos but much of it has been identified and removed in the past. Asbestos is a known cancer-causing material with the greatest risk coming from inhaling the fibers. The government has made funds available for asbestos removal and if a facility manager suspects asbestos has been used in insulating materials, the state agency for managing asbestos removal will be able to assist by providing training and funds for asbestos removal.

If a facility has asbestos insulation, steps should be taken immediately to keep insulation from becoming airborne. This can be accomplished by encapsulating the insulating materials in plastic or another suitable cover, but these types of activities should be conducted by properly trained personnel to minimize the risk to the workers as well as the risk of lawsuits to the facility.

Fiberglass, similar to asbestos but not hazardous, can be encapsulated to prevent loose fibers drifting in the air of the facility.

Pipe hangers and supports should hold the insulation, and not contact the hot water pipe directly.

SAFETY

The facility should include the necessary safety items on a hot water system to protect both the facility and the workers. Of the greatest significance with hot water tanks are the requirements for pressure and temperature relief if the heating equipment should break down and overheat the water.

Pressure and temperature relief devices must be the proper size. An undersized relief device does not allow the hot water out of the tank quickly enough. The risk from an undersized pressure relief device is just as great as if the device were not installed.

Water Heating Safety Standards

For water heating equipment, several industries study and recommend safe practices and standards. The American Society of Mechanical Engineers publishes the Boiler and Pressure Vessel Code. Most safety devices are tested and will have a label or sticker on them. Most common symbols are the UL labels provided by Underwriters Laboratories, Inc. The manufacturer submits his product to, and in most cases pays for, its equipment to be tested under rigorous conditions. The tests are usually required by the manufacturer's insurance company to protect manufacturer's liability. The facility manager should be aware that hot water relief devices and thermostatic controls of burners are subject to these requirements. If the facility does not use tested and approved devices, the risk and liability of the facility is increased dramatically.

Temperature Control

In addition to pressure and temperature relief, electrical instruments are used to control the heating elements. These devices should automatically shut off the heat if temperatures or pressures are exceeded.

VALVES

Hot water system valves are the same as the ones used for cold water, and valves should be provided to isolate sections for maintenance while other parts can be left on line. For example, hot water recirculating pumps are duplicated in facilities where hot water use is critical—e.g., hospitals. Valves should be installed to isolate one pump for maintenance service while the second one provides hot water throughout the facility.

INSTRUMENTS

Pressure and temperature gauges are usually provided on large hot water tanks. In addition, fuel pressure gauges and flowmeters are sometimes installed. Recent electronic trends have led manufacturers to provide computers that record the uses of hot water. The most sophisticated of these programs "remembers" when hot water is used and adjusts the thermostats accordingly. Computers can set the thermostats back on nights or weekends to conserve energy and save hot water heating costs.

A good example of this type of system might be a school where hot water is needed during the week for meals and students' showering needs. On the weekend, however, hot water use is curtailed because the use of hot water is reduced. The computer program "remembers" that on Monday school starts again and begins bringing the temperature up late Sunday evening—when school begins, the entire hot water system is ready for a busy day.

MAINTENANCE

The facility manager should make sure the hot water system is checked on a regular basis. Most manufacturers will furnish a recommended maintenance schedule for their equipment. Routine checks should be recorded both to document the work and to provide trends. Many systems or parts of systems are duplicated to reduce downtime. Weekends and holidays can be used to schedule major repairs provided craft can be scheduled to work on those days.

Another method of providing hot water during a major maintenance activity is contracting for hot water services from a vendor. A temporary tank and heater is set up and water lines are run from inside the facility out to the temporary equipment where it is heated. The hot water is then tied back into the system downstream from the hot water equipment needing work.

Deadly Bath Draws VA Scrutiny

Nurses and Maintenance Faulted in Scalding Death

By Chet Bridger
Federal Times Staff Writer

The scalding death of a patient at the West Los Angeles Veterans Affairs Medical Center has VA officials examining the responsibilities of nurses and building maintenance workers.

Thomas O'Neil, a 45-year-old, brain-damaged patient, died in February after being scalded in 160-degree water.

A department investigation identified several performance issues including:

• A nurse assistant left O'Neil alone in the tub for five minutes or more.

• The psychiatric nursing staff failed to diagnose a non-psychiatric health emergency, the extent of O'Neil's burns.

• The hot water system was not properly maintained and inspected by the maintenance department.

Undersecretary for Health Dr. Kenneth Kizer has ordered all VA hospitals to examine their engineering systems and properly adjust hot water temperatures.

A charge nurse and nursing assistant involved in the incident were removed from patient care pending further review. Kizer retained a panel of outside experts to review O'Neil's care.

O'Neil took the fatal bath about 5:30 p.m. Feb. 7. He was a "total care" patient in the hospital's 15-bed, locked psychiatric care unit.

A nursing assistant drew the bath and tested the water. Two staff members typically bathe total care patients, but a staff shortage permitted only one nursing assistant for O'Neil, according to the VA report.

O'Neil splashed a lot of water on the floor, soaking linens and dry clothing. The assistant left to get new towels and clothes.

Three nursing assistants were caring that evening for the 15 patients with severe mental problems. The other two were in distant parts of the ward, according to the VA report. The assistant chose to leave O'Neil alone rather than call another assistant or the charge nurse.

"While [the nursing assistant] could have used an alert button in the bathroom, this was generally not used by the staff because of its ineffectiveness," according to the VA report.

"Staff felt that the call lights did not work properly and did not alert staff in places other than the nurses' station."

The nursing assistant returned to find O'Neil had turned on the hot water, which was showering over him.

O'Neil immediately was removed from the bath and the charge nurse examined his burns. O'Neil's vital signs were normal. After the psychiatric doctor on duty arrived, O'Neil was sent to the emergency room at 7:13 p.m. He was moved to a burn treatment unit, where he died a couple days later.

Water in the psychiatric unit was too hot because of several mechanical failures.

A pipefitter foreman and plumber recently had fixed a sewage problem, which required shutting off steam and water in the building.

An inspection after O'Neil's death found a malfunctioning steam valve, an incorrectly adjusted thermostat, an inoperational water mixing valve, and an improperly opened bypass valve, according to the VA report.

"The pipefitter [and] plumber noted that the mixing valves throughout the VA North Campus are old and that they are not assigned any regularly scheduled maintenance. They noted that these systems are only fixed when broken," the report said.

Lynna Smith, president of the Nurses Organization of Veterans Affairs, would not comment specifically on the incident, but said it generally is not good practice to leave a total care patient alone in a tub. But she noted that many VA hospitals don't have enough nurses.

> *Water in the psychiatric unit was too hot because of several mechanical failures.*

Figure 12-3. News article "Deadly Bath Draws VA Scrutiny." Courtesy: ©1995, *The Federal Times Newsletter*.

Case Study: Water Too Hot of VA Hospital

The newspaper article shown in Figure 12-3 indicates what can happen to a facility that fails to manage its hot water supply system property. The article relates the death of a mental patient who was left alone in a bathtub for a few minutes. While alone, the patient turned on the hot water shower, scalding himself. The burns were extensive and the patient died a few days later. As can be seen from the attached article, infighting between the hospital nursing staff and the facility maintenance personnel exists as to whether the hot water system was properly maintained.

Hospitals have standards for maintaining hot water temperatures at about 105 degrees to prevent scalding of patients. The latter part of the article indicates several maintenance problems that are discussed here item by item:

1. "A pipefitter foreman and plumber had recently fixed a sewage problem which required shutting off steam and water in the building."

This would be normal. Systems have to be periodically shut down for maintenance. Procedures usually require the staffs to be notified when systems are going to be shut, off to enable the staff to make the necessary adjustments to their plans. Often, however, staff notification means notifying the supervisor. If the supervisor in turn does not inform his own staff, the effect of notification is lost.

2. "An inspection after the patient death found a malfunctioning steam valve..."

Again, not a disastrous condition by itself. Steam valves and other types of valves malfunction and, provided the maintenance staff is trained, the system can continue to operate until maintenance is scheduled.

3. "...an incorrectly adjusted thermostat..."

Depending upon the circumstances, this would be highly irregular in a hospital caring for mental patients, The articles does not indicate, however, if this were the only use. For example, if the kitchen were served with this same water, the thermostat could have been set high for food preparation or dishwashing. In general, management should have been informed when hot water thermostat temperatures were raised. It is likely that a logbook may record the setting, time and date when it was changed.

Facility managers should make sure that accurate records are kept, that the maintenance staff recognizes the implications of deviating from established norms, notifies higher management when they do, and has the authority to make the changes.

4. "...an inoperational water mixing valve..."

The maintenance staff may have not have been aware that this valve was not operating until after the event. In this situation, the maintenance staff has to rely upon the user to call and notify them that the water is too hot. Since the incident occurred at 5:30 p.m., it is likely that shifts had changed and the incoming staff was not aware of the current condition of the system.

5. "...and an improperly opened bypass valve."

The article does not say what was bypassed. Since the statement was made in the official report and it sounds bad, the article writer has repeated it here. The lesson learned from these six words for the facility manager really is: Be careful what is put into a report because people who do not know what they are writing about will read them.

6. Finally, the article indicates that "mixing valves, were only fixed when broken," implying that there may not have been a routine program for maintenance. However, journalistic license may be at work because routine maintenance could be as simple as checking the hot water at the faucets.

If a facility manager is not aware of it already, there are lots of opportunities for second-guessing after the fact. The Veterans Administration has the unfortunate position of having each event that occurs broadcast for the nation to see. Facilities should learn what they can from a tragic event or an upset condition, and get on with business as quickly and quietly as possible.

Chapter 13

Wastewater Systems

Water in, water out... The facility can treat or partially treat its wastewaters or it can work in close coordination with the utility to make sure its wastes are treated safety. The facility manager does not often need to treat raw sewage. The purpose of this chapter is to provide a basic knowledge that is used by city and county water treatment managers, and cover the fundamentals of treating the facility's own wastewater.

SEWAGE TREATMENT

Many facility managers find it necessary to treat their own domestic sewage, while large industrial plants are required to treat sewage.

In addition, recent new trends in water conservation and consumption have generated an interest in "graywater" systems. In a graywater system, the mild waste water from washing hands, bathing, and kitchen waste is reused before being sent to the sewage treatment plant.

The Three-Step Sewage Treatment Process

Sewage treatment is a simple three-step process that makes the most use of natural microbes to decompose and break down human and animal waste. To put it simply, there are bigger bugs out there that like to eat the dangerous pathogenic organisms that contaminate water. A wastewater treatment plant makes it easy for these natural predators to thrive.

As discussed in Chapter 4, the primary mechanism for serious disease transmission is the fecal/oral route. Wastes from humans, including disease pathogens is inadvertently ingested by other humans who become infected and in turn add more pathogens to wastewaters.

To prevent the spread of disease, the wastewater system carries water from toilets, wash basins, sinks, tubs and floor drains to where it can be treated in a central facility. Wastewaters are carried from the facility in pipes. While Chapter 10 provides a fairly detailed discussion of hydraulics or pipe sizing theory, Table 13-1 provides some quick wastewater pipe-sizing guides from the plumbing codes.

The Two Types of Wastewater Systems

The main types of wastewaters include stormwater and sanitary sewer. Stormwater is the runoff from roofs, gutters, downspouts, parking lots, etc. Sanitary sewers carry the water that goes down the drain or down the toilet. They carry waste away from where people are. A new trend is to separate the sanitary waste into graywater from sinks and tubs and blackwater from the toilet and the kitchen garbage disposal unit. The two types of wastes can be conveyed to different areas. In general, the intent is to use graywater for irrigation, plants, fountains and the like—where it is not used for drinking, but it is also not wasted at the central treatment plant.

Many areas still treat the two systems as one sanitary system.

Wastes are carried through pipes to a central wastewater facility. For many facility managers, this will be the city utility and the facility has little responsibility. For others, the wastes are disposed of on site.

Permits

As discussed in Chapter 3 all wastewater treatment facilities are required to have some sort of permit to release the water from the wastewater treatment system back into the environment. Even lagoons, which would be thought to hold water until it evaporates, are required to have a groundwater discharge permit.

The permit regulates the quantity and the quality of the release and provides requirements for monitoring to assure that the releases do not contaminate the environment.

Table 13-1. Estimated Waste/Sewage Flow Rates.

Because of the many variables encountered, it is not possible to set absolute values for waste/sewage flow rates for all situations. The designer should evaluate each situation and, if figures in this table need modification, they should be made with the concurrence of the Administrative Authority.

Type of Occupancy	Gallons (liters) Per Day
1. Airports	15 (56.8) per employee
	5 (18.9) per passenger
2. Auto washers	Check with equipment manufacturer
3. Bowling alleys (snack bar only)	75 (283.9) per lane
4. Camps:	
Campground with central comfort station	35 (132.5) per person
Campground with flush toilets, no showers	25 (94.6) per person
Day camps (no meals served)	15 (56.8) per person
Summer and seasonal	50 (189.3) per person
5. Churches (Sanctuary)	5 (18.9) per seat
with kitchen waste	7 (26.5) per seat
6. Dance halls	5 (18.9) per person
7. Factories	
No showers	25 (94.6) per employee
With showers	35 (132.5) per employee
Cafeteria, add	5 (18.9) per employee
8. Hospitals	250 (946.3) per bed
Kitchen waste only	25 (94.6) per bed
Laundry waste only	40 (151.4) per bed
9. Hotels (no kitchen waste)	60 (227.1) per bed (2 person)
10. Institutions (Resident)	75 (283.9) per person
Nursing home	125 (473.1) per person
Rest home	125 (473.1) per person
11. Laundries, self-service	
(minimum 10 hours per day)	50 (189.3) per wash cycle
Commercial	Per manufacturer's specifications
12. Motel	50 (189.3) per bed space
with kitchen	60 (227.1) per bed space
13. Offices	20 (75.7) per employee
14. Parks, mobile homes	250 (946.3) per space
picnic parks (toilets only)	20 (75.7) per parking space
recreational vehicles—	
without water hook-up	75 (283.9) per space
with water and sewer hook-up	100 (378.5) per space
15. Restaurants—cafeterias	20 (75.7) per employee
toilet	7 (26.5) per customer
kitchen waste	6 (22.7) per meal

(CONTINUED)

Table 13-1. Estimated Waste/Sewage Flow Rates (*Continued*).

```
        add for garbage disposal ............................................................. 1 (3.8) per meal
        add for cocktail lounge ................................................... 2 (7.6) per customer
        kitchen waste—
        disposable service ........................................................... 2 (7.6) per meal
16.  Schools—Staff and office ................................................ 20 (75.7) per person
        Elementary students ............................................... 15 (56.8) per person
        Intermediate and high ..............................................20 (75.7) per student
            with gym and showers, add ...............................................5 (18.9) per student
            with cafeteria, add ............................................................3 (11.4) per student
        Boarding, total waste ............................................ 100 (378.5) per person
17.  Service station, toilets.............................................. 1000 (3785) for 1st bay
                                              500 (1892.5) for each additional bay
18.  Stores .............................................................................20 (75.7) per employee
        public restrooms, add ............................... 1 per 10 sq. ft. (4.1/M²) of floor space
19.  Swimming pools, public ................................................. 10 (37.9) per person
20.  Theaters, auditoriums ...................................................... 5 (18.9) per seat
        drive-in ........................................................................ 10 (37.9) per space
```

(a) **Recommended Design Criteria**. Sewage disposal systems sized using the estimated waste/sewage flow rates should be calculated as follows:

(1) Waste/sewage flow, up to 1500 gallons/day (5677.5 L/day) Flow × 1.5 = septic tank size

(2) Waste/sewage flow, over 1500 gallons/day (5677.5 L/day) Flow x 0.75 + 1125 = septic tank size

(3) Secondary system shall be sized for total flow per 24 hours.

(b) Also see Section K 2 of this appendix.

WASTEWATER TREATMENT METHODS

In the past, stormwater and sanitary sewer water were combined but the huge inflow of water to the treatment plant during a rainstorm upset the balance of water chemistry. Often, the stormwaters overflowed the main treatment plant, causing operators to have to dump their raw sewage into the downstream river or lake. As a result, the system designs were changed and the sanitary sewer system was isolated from the stormwater system.

Today, the sanitary sewer manholes are solid, while in the past these manhole covers were grated.

Sanitary sewer water is conveyed to a central treatment plant. The amount of wastes in the water is small—perhaps only as much as one-tenth of one percent, while the rest of the water acts as conveyance to carry the small amount of wastes.

Treatment is broken down into two or three components. Primary treatment is where the large particles, grease, paper and debris is separated. Secondary treatment is where microbes are allowed to process the wastes. In a few rare instances, tertiary treatment refines and further purifies the water. Tertiary treatment is expensive and is not required except in a few pristine areas where the local public has decided to take the extra steps.

Treatment of sewage wastes then consists of removing the larger particles, providing a location of the microbes to break-down the wastes, and final purification and clarification.

Biochemical Oxygen Demand

The main characteristic for microbe measurement is called biochemical oxygen demand (BOD, pronounced "Bee Oh Dee"). BOD is used in water treatment studies to measure the presence or number of microorganisms in the water.

BOD is actually the measurement of the amount of dissolved oxygen in a water sample and it represents the amount of wastes that can be consumed by microbes. BOD is usually expressed in milligrams of oxygen per liter of water. As the organic matter, which is what is represented by the BOD, is consumed by the bacteria, the bacteria "breathe" the dissolved oxygen in the water. The demand is representative of the microbes use of the oxygen. The bacteria will grow as necessary to consume the organic matter in the sanitary sewer water.

One of two things can happen to the water as the bacteria colony grows and consumes the organic matter. If enough oxygen is present, the bacteria will consume all of the organic matter. After consuming all the organic material, the bacteria, having no further food source will die and sink to the bottom of the vessel. This residual is known as sludge. If there is not enough oxygen in the water to consume the all of the organic matter, a new kind of bacteria that does not need oxygen begins to grow.

The bacteria that grow in oxygen-starved water are called anaerobic bacteria (growing without oxygen). Anaerobic bacteria can process more concentrated wastes and do not need light or air to work. However, the by-products of their consuming of the organic matter include methane and hydrogen sulfide gas. Hydrogen sulfide is a foul-smelling, unpleasant odor, while methane is flammable. Both require large amounts of fresh air to dilute the gases to safe and acceptable levels. The gases that are a product of anaerobic bacteria are why sewage treatment plants are remotely located.

Finally, facility managers should know that hydrogen sulfide gas is insidious. At low levels, it is malodorous, but the nasal passages become desensitized to it at low levels. This accounts for not being able to smell the foul odors after just a few moments.

Primary Treatment

At the treatment plant, primary treatment consists of filtering out large particles, dirt, stones, paper and bits of plastic. The primary design requires the flow to slow down enough so that heavy material can settle to the bottom. In addition, any floating debris is removed. Grinder pumps chop and break up most large pieces of debris. The sludge that settles is removed periodically.

Secondary Treatment

Secondary treatment is essentially the biological process and there are two basic methods for removal of the organic matter. Both methods provide air and microorganisms to interact and reduce the solids into sludge, carbon dioxide and water. The two methods mentioned here are trickling filters and activated sludge.

TRICKLING FILTERS

For most medium-sized communities, a trickling filter system is used. The trickling filter sprays sewage, after it has been through primary treatment, onto a bed of loose rocks or in some cases plastic saddles similar in surface area to rocks. Microorganisms form a slime layer on the rocks and as the sewage trickles over them, the large microorganisms consume the organic matter.

After the water passes through the rock layer, the treated water is Collected at the bottom and is released to ponds, rivers or streams. Trickling filters are usually round, 30-80 ft. in diameter and require pumps to push the sewage through the sprinkling mechanism (see Figure 13-1).

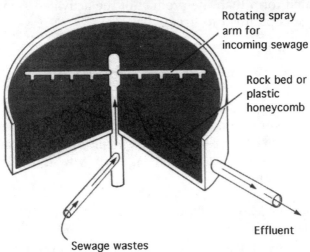

Rotating spray arm for incoming sewage

Rock bed or plastic honeycomb

Effluent

Sewage wastes

Figure 13-1. Typical trickling filter for treatment of wastewater. Reprinted (with minor adaptation) from Microbiology. An introduction with permission of Benjamin/Cummings Publishing Co., Inc.

ACTIVATED SLUDGE

Like the trickling filter process, the activated sludge process uses air to encourage microbes to grow and consume organic matter in raw sewage. After primary treatment, mixed sewer water and activated sludge is pumped into a tank where compressed air is bubbled up through the sludge. The air feeds the microorganisms which consume the organic matter. As the matter is consumed, sludge settles to the bottom of the tank where it is siphoned off to drying beds or to a second anaerobic process.

ANAEROBIC DIGESTER

The activated sludge, in the form of a slurry, flows into another large tank called an anaerobic digester. In this tank the growth of anaerobic (non-oxygen) bacteria is encouraged. These microbes consume the remaining organic matter. The remaining water is released to lagoons to evaporate, or to rivers or streams. The other by-product of the anaerobic digester—sludge is a more concentrated form than the product of the activated sludge tank. The sludge is pumped to drying beds.

Sludge

After drying, the sludge is picked up using earth-moving equipment. Front-end loaders scoop up the dried sludge where it is placed in dump trucks. The trucks can haul the dried sludge to a disposal site or in some areas of the country, the sludge is incinerated. It can even be recycled for agricultural use because it is rich in humus, a product that aids in plant growth (if allowed by the local government).

Problems of Secondary Treatment

The trickling filter, the activated sludge digester and even to some extent the anaerobic digester are subject to poisoning of the microbes by strong chemicals. In effect, the chemicals kill or severely reduce the life of the microorganisms. When this happens, the treatment plant has no choice but to store the sewage until a new batch of microbes is ready. For this reason, many treatment facilities have more than one of these systems. In addition, the tanks and filters have to be taken down periodically for maintenance and for cleaning.

OTHER SEWER TREATMENT FACILITIES

Wastewater may be treated in a number of other facilities, including lagoons, septic tanks, storage tanks and others.

Lagoons

For many small facilities, simple sewer lagoons provide the necessary treatment. The raw sewage flows through grinders to chew up the solids. The remaining effluent flows into a pond where aerators stir the water and mix in air. The organic matter is removed via the microbial action discussed above. From the lagoon, wastewater is held until it evaporates, or it flows through a successive series of ponds until it is clean enough to be released back into rivers, a lake or the ocean. By comparison, lagoons are the simplest to operate but require much more room than either trickling filters or activated sludge digester.

Septic Tanks

For small residential or business users located a considerable distance from a sewer main or in an area that is not served by a central treatment plant, a septic tank is a simple, economical waste treatment alternative provided it is allowed by the regulating agency.

A septic tank is essentially an on-site disposal process that uses anaerobic bacteria. The septic tank is a vault with a dividing wall. Septic tanks can be made from concrete, fiberglass or other suitable leak proof materials (see Figure 13-2). The septic tank is buried sufficiently far from the facility to provide a margin of safety from any fumes coming from the tank and to prevent leaching back from the septic tank into the facility. Recommended distances for building sewers and septic tanks are provided in Table 13-2.

Raw sewage flows into the first portion of the septic tank as shown in the figure. Here, the large heavy solids drop out, and any grease or other floating debris is skimmed. In the second part, the anaerobic process takes place and the organic matter in the sewage is consumed.

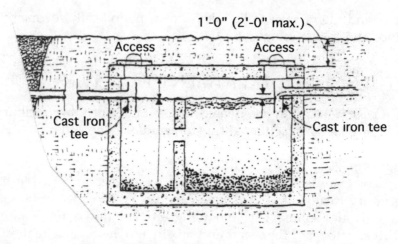

Figure 13-2. Typical septic tank. Courtesy: Stein and Reynolds, *Mechanical and Electrical Equipment for Buildings,* **8th Edition. ©1992 John Wiley & Sons, reprinted with permission.**

From the septic tank, the water or *effluent* flows to a leach field. The septic tank leach field is a series of slotted or holed pipes where the waste water flows. The released water irrigates sod or other vegetation. Usually, the leach field does not supply vegetable gardens because on occasion the decomposition process is not completed in the septic tank. The sizing and function of the drain field is dependent upon the soil conditions in the area.

Septic tanks work well for single-family dwellings and small remote businesses but are easily overloaded, and the process can be interrupted by oil, strong bleach or chemical cleaning compounds, or other chemicals that are toxic to the microbes. In addition, large amounts of sand, grease or garbage can overload the primary chamber and in these instances an interceptor is required.

Every few years, septic tanks must be pumped out to remove debris and solids that have filled up the primary vessel.

Storage Tanks

One short-term facility solution to sewerage disposal problems is to provide an underground storage tank where waste waters can flow. Then, periodically, the facility hires a septic tank pumping company to come and pump the sewage and haul it in

Table 13-2. Location of Sewage Disposal System.

Reproduced from the 2000 edition of the Uniform Plumbing Code™, ©1999, with permission of the publishers, the International Association of Plumbing and Mechanical Officials. All rights reserved.

Minimum Horizontal Distance In Clear Required From:	Building Sewer		Septic Tank		Disposal Field		Seepage Pit or Cesspool	
Buildings or structures[1]	2 feet	(610 mm)	5 feet	(1524 mm)	8 feet	(2438 mm)	8 feet	(2438 mm)
Property line adjoining private property	Clear[2]		5 feet	(1524 mm)	5 feet	(1524 mm)	8 feet	(2438 mm)
Water supply wells	50 feet[3]	(15240 mm)	50 feet	(15240 mm)	100 feet	(30.5 m)	150 feet	(45.7 m)
Streams	50 feet	(15240 mm)	50 feet	(15240 mm)	50 feet[7]	(15240 mm),	100 feet[7]	(30.5 m)[7]
Trees	—		10 feet	(3048 mm)	—		10 feet	(3048 mm)
Seepage pits or cesspools	—		5 feet	(1524 mm)	5 feet	(1524 mm)	12 feet	(3658 mm)
Disposal field	—		5 feet	(1524 mm)	4 feet[4]	(1219 mm)	5 feet	(1524 mm)
On site domestic water service line	1 foot[5]	(305 mm)	5 feet	(1524 mm)	5 feet	(1524 mm)	5 feet	(1524 mm)
Distribution box	—		—		5 feet	(1524 mm)	5 feet	(1524 mm)
Pressure public water main	10 feet[6]	(3048 mm)	10 feet	(3048 mm)	10 feet	(3048 mm)	10 feet	(3048 mm)

Note:
When disposal fields and/or seepage pits are installed in sloping ground, the minimum horizontal distance between any part of the leaching system and ground surface shall be fifteen (15) feet (4572 mm).

1. Including porches and steps, whether covered or uncovered, breezeways, roofed porte-cocheres, roofed patios, carports, covered walks, covered driveways and similar structures or appurtenances.
2. See also Section 313.3 of the Uniform Plumbing Code.
3. All drainage piping shall clear domestic water supply wells by at least fifty (50) feet (15240 mm). This distance may be reduced to not less than twenty-five (25) feet (7620 mm) when the drainage piping is constructed of materials approved for use within a building.
4. Plus two (2) feet (610 mm) for each additional (1) foot (305 mm) of depth in excess of one (1) foot (305 mm) below the bottom of the drain line. (See also Section K 6.)
5. See Section 720.0 of the Uniform Plumbing Code.
6. For parallel construction—For crossings, approval by the Health Department shall be required.
7. These minimum clear horizontal distances shall also apply between disposal field, seepage pits, and the ocean mean higher high fide line.

trucks to a treatment disposal location. As regulations increase relative to the disposal of sanitary wastes, this option will become more and more attractive. It is often used as a temporary measure during construction of a large facility while the sanitary sewer system is still under construction. Underground sewer tanks are regulated by the County Health Department and are leak-tested and inspected.

In addition, concrete blocks are placed on the tank's sides and straps placed over it to hold it should a large rainstorm or tidewater run into the soil around the tank, making it try to float out of the ground. The lifting force can be several tons for a 5,000 gallon-tank.

Other Types/Packaged Units

Several vendors make packaged units available that are combinations of the sewage disposal types discussed here. A method of evaluating them would be a combination of cost, BOD removal, size and utility requirements.

Oxygenated Ditch

A small effective sewer treatment facility is known as the oxygenated ditch, which uses a oval or circular pattern where the raw sewage flows after primary treatment. In this system, a paddle-wheel moves the water in the ditch. The paddles inject air. The movement agitates the flow to allow mixing and aeration of organic matter. Baffles along the bottom stop the sludge and augers move the sludge into the center where it is pumped into drying beds. If necessary, the clear effluent can be drawn off, but usually the water is allowed to evaporate. The oxygenated ditch takes more room than trickling filters, but less than lagoons. Its advantage is that it requires few operators and is fairly forgiving if injected with chemicals.

SOIL CONDITIONS

The design of sewage lagoons, septic tank leach fields, and to some extent buried pipe are dependent upon the soil conditions in the area. A detailed discussion of soil types and parameters is beyond the scope of this text, but a few brief items are mentioned here for the facility manager.

Rock

Rock is ideal subgrade material—however, removing it is expensive. If there is a choice, do not use rock for this purpose.

Sand and Gravel

Sand and gravel is excellent material and it is highly porous. Most codes do not allow constructing sewage lagoons on sand or gravel because the high porosity will allow the sewer discharges to leach to groundwater.

Clay

There are two kinds of clays, called fat clays and lean clays. Fat clays form a gummy sticky material when a clay is wet. These clays are excellent barriers for drainage when thick enough. Most ponds are lined with some form of a fat clay. Lean clays are easier to work with because they are also impervious but do not ball up as much. Lean clays bind to themselves better than fat clays. A fat clay will often dry up and blow away while a lean clay will not.

By rolling a ball of wet soil between the palms, a quick judgment can be made between lean and fat clays. If the material rolls out very thin, smaller than a No. 2 pencil lead, then it is a fat clay. If it will not roll much smaller than a pencil before breaking, then it is a lean clay. Often, soil is a combination of these.

Cobble Rock

Occasionally, a lean clay will have cobble rock in it. Cobble rock are stones between 4 and 9 inches. Large stones are not a problem for a pond liner but they provide a path of water along the rock—as a result the liner would need to be thicker.

Cobble rock is not good bedding material for pipes in the ground because the soil will collapse leaving only the stones. With the weight of material over the pipe, the stone can puncture or crack the pipe. Table 13-3 shows septic and leaching rates for five types of soils.

Groundwater

Soil types are affected by groundwater and leaching sewage into the groundwater table is allowed only by permission. Note: Some areas of high groundwater will cause septic and storage type tanks to float if they are water-tight and there is less water in the water tank than in the adjacent ground.

Table 13-3. Design Criteria of Five Typical Soils.

Reproduced from the 2000 edition of the Uniform Plumbing Code™, ©1999, with permission of the publishers, the International Association of Plumbing and Mechanical Officials. All rights reserved.

Type of Soil	Required sq. ft. of leaching area/ 100 gals. (m²/L)		Maximum absorption capacity in gals./sq. ft. of leaching area for a 24 hr. period (L/m²)	
Coarse sand or gravel	20	(0.005)	5.0	(203.7)
Fine sand	25	(0.006)	4.0	(162.9)
Sandy loam or sandy clay	40	(0.010)	2.5	(101.8)
Clay with considerable sand or gravel	90	(0.022)	1.1	(44.8)
Clay with small amount of sand or gravel	120	(0.030)	0.8	(32.6)

Chapter 14
Putting It All Together: Applications

his is where we get to the "nitty gritty" of water system management, by individual application. Recognizing that a well-designed system is less costly to operate and maintain, and armed with the proper knowledge, the facility manager and his team are ready to examine the facility's water-using areas on an existing floor plan or blueprint of a future facility—and begin making quality and cost enhancements.

COMBINING SYSTEMS

Now that the facility manager is more conscious of the many individual elements that make up his water system, he is ready to begin combining some elements into performing the central focus of facility managers—providing for the interface of where people and water systems interact. Initially, the discussion will be for bathrooms and restrooms since this is one of the more common areas of interface that a facility manager must manage. Following the discussion on bathroom/restroom facilities are similar discussions for kitchens, mechanical rooms, swimming and bathing facilities, fountains and spas, clinics and laboratories.

Lavatories
Almost every facility has some type of restroom, whether it be a small one-person unit or a large facility for a ballpark, sports arena, church or office complex. The restroom is going to have water closets and sinks. In addition, the men's room will have urinals.

Codes and Standards for Various Facilities

Hospitals	Joint Commission on Accreditation of Healthcare Organizations
Swimming Pools	National Swimming Pool and Spa Institute
Medical Clinics	American Medical Association
Kitchens	American Institute of Architects
Restrooms	American Institute of Architects, National Sanitation Foundation

The layout of the restroom is largely a matter of design with many human factors incorporated. Some of these factors are driven by the fixtures, some by the plumbing, and still others by less tangible factors.

Table 14-1 shows the approximate number of plumbing fixtures needed based upon the estimated number of facility users for office and public buildings. Each fixture requires enough water supply to operate correctly.

Water Closets and Urinals

The water closets and urinals can have tank flush or flush valves. In general, the tank type has problems in a public facility and are not recommended. For users, the seat should be smooth and comfortable and made from plastic or other nonporous material that will warm quickly. Metal seats, obviously, are not practical.

As discussed in Chapter 9, water closets can be either wall- or floor-mounted. If feasible, the water closets should be wall-mounted which aids in cleaning the bath facility by cleaning personnel.

Each water closet in a public restroom is enclosed by wall panels and a simple lock is fixed to the door. Inside, there should be a hook for a jacket, usually affixed to the inside of the door but not necessary—and, of course, hangers for toilet paper, usually two.

Table 14-1. Minimum Plumbing Facilities[1]

Each building shall be provided with sanitary facilities, Including provisions for the physically handicapped as prescribed by the Department having jurisdiction. For requirements for the handicapped, ANSI A117.1-1992, Accessible and Usable Buildings and Facilities, may be used.

The total occupant load shall be determined by minimum exiting requirements. The minimum number of fixtures shall be calculated at fifty (50) percent male and fifty (50) percent female based on the total occupant load.

Reproduced from the 2000 edition of the Uniform Plumbing Code™, ©1999, with permission of the publishers, the International Association of Plumbing and Mechanical Officials. All rights reserved.

Type of Building or occupancy[2]	Water Closets[3,4] (Fixtures per Person)		Urinals[5,10] (Fixtures per Person)	Lavatories (Fixtures per Person)		Bathtubs or Showers (Fixtures per Person)	Drinking Fountains[3,13] (Fixtures per Person)
Assembly Places—Theatres, Auditoriums, Convention Halls, etc.—for permanent employee use	Male 1:1-15 2:16-35 3:36-55 Over 55, add 1 fixture for each additional 40 persons.	Female 1:1-15 3:16-35 4:36-55	Male 0:1-9 1:10-50 Add one fixture for each additional 50 males.	Male 1 per 40	Female 1 per 40		
Assembly Places—Theatres, Auditoriums, Convention Halls, etc.—for public use	Male 1:1-100 2:101-200 3:201-400 Over 400, add one fixture for each additional 500 males and 1 for each additional 125 females.	Female 3:1-50 4:51-100 8:101-200 11:201-400	Male 1:1-100 2:101-200 3:201-400 4:401-600 Over 600 add 1 fixture for each additional 300 males.	Male 1:1-200 2:201-400 3:401-750 Over 750, add one fixture for each additional 500 persons.	Female 1:1-200 2:201-400 3:401-750		1:1-150 2:151-400 3:401-750 Over 750, add one fixture for each additional 500 persons.
Dormitories[9] School or Labor	Male 1 per 10 Add 1 fixture for each additional 25 males over 10) and 1 for each 20 females (over 8).	Female 1 per 8	Male 1 per 25 Over 150, add 1 fixture for each additional 50 males	Male 1 per 12 Over 12 add one fixture for each additional 20 males and 1 for each 15 additional females.	Female 1 per 12	1 per 8 For females, add 1 bathtub per 30. Over 150, add 1 per 20.	1 per 150[12]

Table 14-1. Minimum Plumbing Facilities[1] (Continued)

Type of Building or occupancy[2]	Water Closets[14] (Fixtures per Person)		Urinals[5,10] (Fixtures per Person)	Lavatories (Fixtures per Person)		Bathtubs or Showers (Fixtures per Person)	Drinking Fountains[3,13] (Fixtures per Person)
Dormitories— for staff use	Male 1:1-15 2:16-35 3:36-56 Over 55, add 1 fixture for each additional 40 persons.	Female 1:1-15 3:16-35 4:36-55	Male 1 per 50	Male 1 per 40	Female 1 per 40	1 per 8	
Dwellings[4] Single Dwelling Multiple Dwelling or Apartment House	1 per dwelling 1 per dwelling or apartment unit			1 per dwelling 1 per dwelling or apartment unit		1 per dwelling 1 per dwelling or apartment unit	
Hospital Waiting rooms	1 per room			1 per room		1 per 150[12]	
Hospitals— for employee use	Male 1:1-15 2:16-35 3:36-55 Over 55, add 1 fixture for each additional 40 persons.	Female 1:1-15 3:16-35 4:36-55	Male 0:11-9 1:10-50 Add one fixture for each additional 50 males.	Male 1 per 40	Female 1 per 40		
Hospitals Individual Room Ward Room	1 per room 1 per 8 patients			1 per room 1 per 10 patients		1 per room 1 per 20 patients	1 per 150[12]
Industrial[6] Warehouses Workshops, Foundries and similar establishments— for employee use	Male 1:1-10 2:11-25 3:26-50 4:51-75 5:76-100 Over 100, add 1 fixture for each additional 30 persons	Female 1:1-10 2:11-25 3:26-50 4:51-75 6:76-100		Up to 100, 1 per 10 persons Over 100, 1 per 15 persons[7,8]		1 shower for each 15 persons exposed to excessive heat or to skin contamination with poisonous, infectious, or irritating material	1 per 150[12]

Table 14-1. Minimum Plumbing Facilities[1] *(Continued)*

Type of Building or occupancy[2]	Water Closets[14] (Fixtures per Person)	Urinals[5,10] (Fixtures per Person)	Lavatories (Fixtures per Person)	Bathtubs or Showers (Fixtures per Person)	Drinking Fountains[3,13] (Fixtures per Person)
Institutional - Other than Hospitals or Penal Institutions (on each occupied floor)	Male 1 per 25 Female 1 per 20	Male 0:1-9 1:10-50 Add one fixture for each additional 50 males.	Male 1 per 10 Female 1 per 10	1 per 8	1 per 150[12]
Institutional - Other than Hospitals or Penal Institutions (on each occupied floor)—for employee use	Male 1:1-15 2:16-35 3:36-55 Over 55, add 1 fixture for each additional 40 persons. Female 1:1-15 3:16-35 4:36-55	Male 0:1-9 1:10-50 Add one fixture for each additional 50 males.	Male 1 per 40 Female 1 per 40	1 per 8	1 per 150[12]
Office or Public Buildings	Male 1:1-100 2:101-200 3:201-400 Over 400, add one fixture for each additional 500 males and 1 for each additional 150 females. Female 3:1-50 4:51-100 8:101-200 11:201-400	Male 1:1-100 2:101-200 3:201-400 4:401-600 Over 600 add 1 fixture for each additional 300 males.	Male 1:1-200 2:201-400 3:401-750 Over 750, add one fixture for each additional 500 persons. Female 1:1-200 2:201-400 3:401-750		1 per 150[12]
Office or Public Buildings—for employee use	Male 1:1-15 2:16-35 3:36-55 Over 55, add 1 fixture for each additional 40 persons. Female 1:1-15 3:16-35 4:36-55	Male 0:1-9 1:10-50 Add one fixture for each additional 50 males.	Male 1 per 40 Female 1 per 40		
Penal Institutions—for employee use	Male 1:1-15 2:16-35 3:36-55 Over 55, add 1 fixture for each additional 40 persons. Female 1:1-15 3:16-35 4:36-55	Male 0:1-9 1:10-50 Add one fixture for each additional 60 males.	Male 1 per 40 Female 1 per 40		1 per 150[12]

Table 14-1. Minimum Plumbing Facilities[1] (Continued)

Type of Building or occupancy[2]	Water Closets[14] (Fixtures per Person)	Urinals[5,10] (Fixtures per Person)	Lavatories (Fixtures per Person)	Bathtubs or Showers (Fixtures per Person)	Drinking Fountains[3,13] (Fixtures per Person)
Penal Institutions— for prison use					
Cell	1 per cell		1 per cell		1 per cell block floor
Exercise Room	1 per exercise room	1 per exercise room	1 per exercise room		1 per exercise room
Restaurants, Pubs and Lounges[11]	Male 1:1-50, 2:51-150, 3:151-300 / Female 1:1-50, 2:51-150, 4:151-300. Over 300, add 1 fixture for each additional 200 persons	Male 1:1-150. Over 150, add 1 fixture for each additional 150 males	Male 1:1-150, 2:151-200, 3:201-400 / Female 1:1-150, 2:151-200, 3:201-400. Over 400, add 1 fixture for each additional 400 persons		
Schools—for staff use					
All schools	Male 1:1-15, 2:16-35, 3:36-55 / Female 1:1-15, 2:16-35, 3:36-55. Over 55, add 1 fixture for each additional 40 persons	Male 1 per 60	Male 1 per 40 / Female 1 per 40		
Schools—for student use					
Nursery	Male 1:1-20, 2:21-50 / Female 1:1-20, 2:21-50. Over 50, add 1 fixture for each additional 50 persons		Male 1:1-25, 2:26-50 / Female 1:1-25, 2:26-50. Over 50, add 1 fixture for each additional 60 persons		1 per 150[12]
Elementary	Male 1 per 30 / Female 1 per 25	Male 1 per 75	Male 1 per 35 / Female 1 per 35		1 per 150[12]
Secondary	Male 1 per 40 / Female 1 per 30	Male 1 per 35	Male 1 per 40 / Female 1 per 40		1 per 150[12]
Others (Colleges, Universities, Adult Centers, etc.)	Male 1 per 40 / Female 1 per 30	Male 1 per 35	Male 1 per 40 / Female 1 per 40		1 per 150[12]

Table 14-1. Minimum Plumbing Facilities[1] (Continued)

Worship Places Educational and Activities Unit	Male 1 per 150	Female 1 per 75	Male 1 per 150	1 per 2 water closets	1 per 150[12]
Worship Places Principal Assembly Piece	Male 1 per 150	Female 1 per 75	Male 1 per 150	1 per 2 water closets	1 per 150[12]

Plumbing Fixtures and Fixture Fittings

1. The figures shown are based upon one (1) fixture being the minimum required for the number of persons indicated or any fraction thereof.
2. Building categories not shown on this table shall be considered separately by the Administrative Authority.
3. Drinking fountains shall not be installed in toilet rooms.
4. Laundry trays. One (1) laundry tray or one (1) automatic washer standpipe for each dwelling unit or one (1) laundry tray or one (1) automatic washer standpipe, or combination thereof, for each twelve (12) apartments. Kitchen sinks, one (1) for each dwelling or apartment unit.
5. For each urinal added in excess of the minimum required, one water closet may be deducted. The number of water closets shall not be reduced to less than two-thirds (2/3) of the minimum requirement.
6. As required by ANSI Z4.1-1968, Sanitation in Places of Employment.
7. Where there is exposure to skin contamination with poisonous, infectious, or irritating materials, provide one (1) lavatory for each five (5) persons.
8. Twenty-four (24) lineal inches (610 mm) of wash sink or eighteen (18) inches (457 mm) of a circular basin, when provided with water outlets for such space, shall be considered equivalent to one (1) lavatory.
9. Laundry trays, one (1) for each fifty (50) persons. Service sinks, one (1) for each hundred (100) persons.
10. General. In applying this schedule of facilities, consideration shall be given to the accessibility of the fixtures. Conformity purely on a numerical basis may not result in an installation suited to the need of the individual establishment. For example, schools should be provided with toilet facilities on each floor having classrooms.
   a. Surrounding materials, wall and floor space to a point two (2) feet (610 mm) in front of urinal lip and four (4) feet (1219 mm) above the floor, and at least two (2) feet (610 mm) to each side of the urinal shall be lined with non-absorbent materials.
   b. Trough urinals shall be prohibited.
11. A restaurant is defined as a business which sells food to be consumed on the premises.
   a. The number of occupants for a drive-in restaurant shall be considered as equal to the number of parking stalls.
   b. Employee toilet facilities shall not be included in the above restaurant requirements. Hand washing facilities shall be available in the kitchen for employees.
12. Where food is consumed indoors, water stations may be substituted for drinking fountains. Offices, or public buildings for use by more than six (6) persons shall have one (1) drinking fountain for the first one hundred fifty (150) persons and one (1) additional fountain for each three hundred (300) persons thereafter.
13. There shall be a minimum of one (1) drinking fountain per occupied floor in schools, theatres, auditoriums, dormitories, offices or public building.
14. The total number of water closets for females shall be at least equal to the total number of water closets and urinals required for males.

Table 14-2. Fixture Unit Table for Determining Water Pipe and Meter Sizes

Reproduced from the 2000 edition of the Uniform Plumbing Code™, ©1999, with permission of the publishers, the International Association of Plumbing and Mechanical Officials. All rights reserved.

Inch	mm
1/2	15
3/4	20
1	25
1-1/4	32
1-1/2	40
2	50
2-1/2	65

Pressure Range - 30 to 45 psi (207 to 310 kPa)

Meter and Street Service, Inches	Building Supply and Branches, Inches	Maximum Allowable Length in Feet (meters)														
		40 (12)	60 (18)	80 (24)	100 (30)	150 (46)	200 (61)	250 (76)	300 (91)	400 (122)	500 (152)	600 (183)	700 (213)	800 (244)	900 (274)	1000 (305)
3/4	1/2***	6	5	4	3	2	1	1	1	0	0	0	0	0	0	0
3/4	3/4	16	16	14	12	9	6	5	5	4	4	3	2	2	2	1
3/4 1		29	23	21	17	15	13	12	10	8	6	6	6	6	6	6
1	1	36	31	27	25	20	17	15	13	12	10	8	6	6	6	6
3/4	1-1/4	36	33	31	28	24	23	21	19	17	16	13	12	12	11	11
1	1-1/4	54	47	42	38	32	28	25	23	19	17	14	12	12	11	11
1-1/2	1-1/4	78	68	57	48	38	32	28	25	21	18	15	12	12	11	11
1	1-1/2	85	84	79	65	56	48	43	38	32	28	26	22	21	20	20
1-1/2	1-1/2	150	124	105	91	70	57	49	45	36	31	26	23	21	20	20
2	1-1/2	151	129	129	110	80	64	53	46	38	32	27	23	21	20	20
1	2	85	85	85	85	85	85	82	80	66	61	57	52	49	46	43
1-1/2	2	220	205	190	176	155	138	127	120	104	85	70	61	57	54	51
2	2	370	327	292	265	217	185	164	147	124	96	70	61	57	54	51
2	2-1/2	445	418	390	370	330	300	280	265	240	220	198	175	158	143	133

Table 14-2. Fixture Unit Table for Determining Water Pipe and Meter Sizes (*Continued*)

Pressure Range - 46 to 60 psi (317 to 414 kPa)**

Meter	Building Supply	Maximum Developed Length (feet) →														
3/4	1/2***	7	7	6	5	4	3	2	2	1	1	1	0	0	0	0
3/4	3/4	20	20	19	17	14	11	9	8	6	5	4	4	3	3	3
3/4	1	39	39	36	33	28	23	21	19	17	14	12	10	9	8	8
1	1	39	39	39	36	30	25	23	20	18	15	12	10	9	8	8
3/4	1-1/4	39	39	39	39	39	39	34	32	27	25	22	19	19	17	16
1	1-1/4	78	78	76	67	52	44	39	36	30	27	24	20	19	17	16
1-1/2	1-1/4	78	78	78	78	66	52	44	39	33	29	24	20	19	17	16
1	1-1/2	85	85	85	85	85	85	80	67	55	49	41	37	34	32	30
1-1/2	1-1/2	151	151	151	151	128	105	90	78	62	52	42	38	35	32	30
2	1-1/2	151	151	151	151	150	117	98	84	67	55	42	38	35	32	30
1-1/2	2	85	85	85	85	85	85	85	85	85	85	85	85	85	83	80
2	2	370	370	340	318	272	240	220	198	170	150	135	123	110	102	94
2	2-1/2	370	370	370	370	368	318	280	250	205	165	142	123	110	102	94
2-1/2	2-1/2	654	640	610	580	535	500	470	440	400	365	335	315	285	267	250

Pressure Range - Over 60 psi (414 kPa)**

Meter	Building Supply	Maximum Developed Length (feet) →														
3/4	1/2***	7	7	7	6	5	4	3	3	2	1	1	1	1	1	0
3/4	3/4	20	20	20	20	17	13	11	10	8	7	6	6	5	4	4
3/4	1	39	39	39	39	35	30	27	24	21	17	14	13	12	12	11
1	1	39	39	39	39	38	32	29	26	22	18	14	13	12	12	11
3/4	1-1/4	39	39	39	39	39	39	39	39	34	28	26	25	23	22	21
1	1-1/4	78	78	78	78	74	62	53	47	39	31	26	25	23	22	21
1-1/2	1-1/4	78	78	78	78	78	74	65	54	43	34	26	25	23	22	21
1	1-1/2	85	85	85	85	85	85	85	85	81	64	51	48	46	43	40
1-1/2	1-1/2	151	151	151	151	151	151	130	113	88	73	51	51	46	43	40
2	1-1/2	151	151	151	151	151	151	142	122	98	82	64	51	46	43	40
1-1/2	2	85	85	85	85	85	85	85	85	85	85	85	85	85	85	85
2	2	370	370	370	370	360	335	305	282	244	212	187	172	153	141	129
2	2-1/2	370	370	370	370	370	370	370	340	288	245	204	172	153	141	129
2-1/2	2-1/2	654	654	654	654	654	650	610	570	510	460	430	404	380	356	329

** Available static pressure after head loss.
*** Building supply, three-quarter (3/4) inch (20 mm) nominal size minimum.

The number of people at the facility determines both the number of water closets or urinals and the number of sinks. Codes require a minimum number of sinks for people to wash their hands after using the restroom. The sink should provide both hot and cold water. When the number of water closets, urinals, and sinks has been estimated, room dimensions are established.

Sinks

Restroom sinks can be free-standing, recessed in a counter or the pedestal type. A hot and cold water faucet along with a drain is provided. Codes specify the distance from the sink drain to the trap and from the trap to the vent. The height of the sink is standardized in the U.S. as is the height for the water closet (see Figure 14-1).

Along with the sink will be a soap and paper towel dispenser, but the trend in recent years has been to locate the paper towel dispenser more than a hand's reach away from the sink to

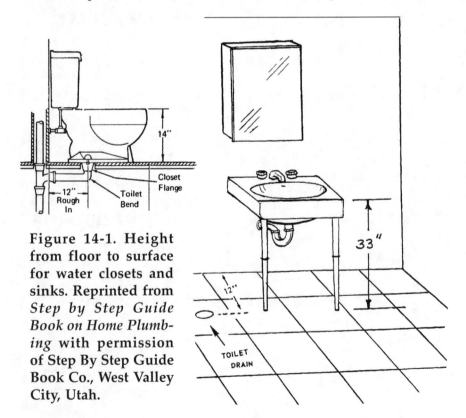

Figure 14-1. Height from floor to surface for water closets and sinks. Reprinted from *Step by Step Guide Book on Home Plumbing* with permission of Step By Step Guide Book Co., West Valley City, Utah.

prevent the paper towels from falling into the sink and plugging it. A warm air dryer is still popularly used for drying hands but a dryer requires electrical wiring power to the units.

Layout

As the architect lays out the facility, the restrooms will be strategically located to provide optimum location. In a large multistory building, for example, a restroom is provided on each floor. Given the size of the floor, there may be more than one restroom on each floor. For some designs, executives may insist on their own private washroom.

The sizes and layouts of these facilities is a matter of design. Knowing what works and what does not is a matter to be determined by architects. While the plumbing codes will provide recommendations for the number of users per water closet and/or sink, general guidance is needed to determine a practical distance to the rest room from the patron's station or the worker's desk, no one standard governs all situations.

Once the restroom has been roughly laid out by the architect, the details of the physical dimensions can be finalized. The wall panels that surround the water closets come in fixed dimensions in increments of two inches. Both the minimum width of the cubicles and the recommended width are also defined by architectural standards. Space should be provided outside of the cubicles for movement and to prevent the patron from becoming trapped at one end of a long narrow corridor.

In 1990, Congress passed the Americans with Disabilities Act (ADA), which mandated requirements to provide restroom facilities for those of us who use a wheelchair or have to use other assistance for walking. Access, grab rails, toilet and sink fixtures are required to accommodate the handicapped under the ADA requirements. For more information about the ADA, see *Interiors Management: A Guide for Facility Managers* by Maggie Smith and *Safety Management: A Guide for Facility Managers* by Joseph Gustin, companion volumes in The Facilities Management Library.

Accessories

In most restroom facilities, the hand dryer is located on a wall at least 8 ft. from the sink. This way, it is physically impossible to touch the sink and the electric dryer element at the same time, reducing the potential for electric shock and reducing the requirement for power and plumbing to be in the same wall.

In women's restrooms, in addition to the paper towel dispenser a sanitary napkin dispenser is provided and restocked by the cleaning personnel.

So for restrooms, in addition to the water closets, urinals and sinks, there are accessory items: wall partitions, toilet paper hangers, hooks, paper towel dispenser(s), soap dispenser(s), soap dish(es), and for females, sanitary napkin dispensers.

On urinals, an infrared position sensor can be fitted. The urinal will automatically flush when the user steps away from the fixture.

The advantage of the infrared device is that it enhances cleanliness because the urinal is flushed each time it is used.

After the decisions have been made about the number of water closets and sinks, decisions should be made about a few other human factors as well. In many areas, a mirror is placed in front of the sink for people who wish to check their appearance after washing. The mirror has to be placed strategically. Many is the facility manager who is embarrassed when discovering that other restroom patrons can clearly be seen relieving themselves by looking at the mirror on the restroom wall!

Lighting is a factor to consider carefully. Lighting should be placed over the mirrors and over the water closets as well. In some facilities, the architect, to save lighting costs, provides light fixtures over the general area and at the sinks, but nothing over the water closets. This mistake leads to dark cubicles around the water closets and prevents the housekeeping staff from seeing clearly enough to assure complete cleaning. This in turn creates doubt about the cleanliness of the water closet, and in general reflects poorly upon the entire complex.

While the extra lighting is a little more costly, it is well worth the expense in cleanliness and the resulting positive attitude it reflects upon the restroom and hence the manager of the facility

Construction Materials

The walls and floor should be made of water-resistant material that is readily cleaned and does not allow for growth of molds or stains. Ceramic tile has been one of the more attractive choices for the past 50 years. Tile, if it is kept clean and maintained, will last a long time, up to 100 years, and certainly outlasts the life of the facility. Flooring can be ceramic tile as well; however, floor tiles usually have sand or other granular material embedded in it to prevent slips and falls. Clay tiles work well along with vinyl flooring. The sink counter tops can be tile or Formica.

Color of the tiles, flooring and ceiling is usually coordinated. A good designer or contractor will prepare a color board for the owner. Here, instead of just passing out a chain with color chips for selection, a board is prepared that shows the different types of tiles, flooring, wall panels, soap, towel and paper dispenser. The owner/facility manager can look over the combinations of colors and textures and make selections based upon the known color chips and tile components. In restrooms, light colors work best because they reflect light and reveal stains. This allows easy inspection for cleanliness. Black, dark brown or dark gray colors should be avoided.

Since the plumbing and vent lines are in the walls, a cleanout should be located in a readily accessible spot, and depending upon the frequency of cleaning, a floor drain might not be a bad idea.

Finally, the facility manager has to work with the architect to decide if the restroom will be on an outside wall. The advantage is for windows and exhaust fans or ducts to be located on the outside wall with a reduction in the amount of air conditioning ducts required. Bathrooms, restrooms and other bathing facilities need a greater exchange of air than rooms like sleeping rooms or offices. The increased air vents noxious fumes associated with restroom facilities. Needless to say, the outside wall that has a fan in it directly to the outside should not be over the business front sidewalk or an outside eating, picnic or rest area.

If a window is used in a bathroom with an outside wall, the glass should be frosted, of course, and only on upper floors. Open windows provide viewing by casual observers. If windows are not provided, emergency lighting should be installed somewhere inside the restroom since, if the power is lost, restrooms without windows become quite dark. The lighting can be a plug-in unit or the light fixtures themselves can have batteries inside them that take over and provide minimal light when the power fails.

Figure 14-2 shows a layout of a restroom fitted with sinks, counter tops, mirrors and lighting.

Kitchens

There are many similarities between construction of restrooms and kitchens, but there are also some large differences. Since the kitchen is the center of a facility for food preparation, its design varies from bathrooms. Most of the water management facilities are essentially the same.

Figure 14-2. A typical restroom layout.

Both hot and cold water supplies are plumbed into the kitchen for washing and preparing food. Drains are provided to carry waste away. In addition, kitchens have heating equipment, such as ovens, grills and stoves for cooking the various dishes. Finally, there is a dishwasher or other types of equipment items for washing and cleaning the dishes themselves. Provision is made for cold storage of foods in the form of a refrigerator or even a walk-in cold storage room.

Central elements of design in a kitchen are the use of non-combustible materials in construction of the shelves, counters, furniture and the use of ceramic or stainless steel to aid in cleaning and sterilizing the food preparation surfaces.

Water is never used over grills, fat fryers or barbecues as water dumped on these types of cooking devices will cause injury from splashing hot grease. Fire protection for grills is a dry chemical system that dumps a coating over the burning material and smothers any fire.

Water fixtures used for cleaning include large, deep sinks for washing pots or pans too large for the dishwasher, and wide shallow sinks for cleaning vegetables and fruits.

The dishwasher is usually integral with the countertops in the area reserved for cleaning dirty dishes and a warm water nozzle is piped overhead for general overall spray and washdown. A large garbage disposer grinds garbage, allowing it to wash down into the sewer system. Most restaurants will have a grease trap in the drain line designed specifically to capture grease which will solidify when combined with the cooler water in the main sewer line (see Figure 14-3).

The grease trap is usually located near the facility, but it is not in the kitchen because cleaning a grease trap is a dirty, messy operation and it cannot be done while meals are being prepared.

Dishwasher

A dishwasher for a large facility is usually hard plumbed into the kitchen. It is supplied with either hot and cold water or it has its own heating unit which heats cold water directly and uses it in the wash cycles. Commercial dishwashers also have a large garbage disposer integral with the drain for grinding any leftover food materials or other small items.

CAST IRON GREASE INTERCEPTORS

FUNCTION: The all steel receiving cradle provides positive support for the interceptor and adjustability for the inlet and outlet connections to meet the drainage piping. The receiving cradle can be set in the slab at the time of the pouring or set in a pre-sleeved opening in the rough slab. The receiver is secured and suspended in the rough slab by its anchoring flange. The interceptor body is lowered into the receiver cradle and adjusted to the connecting drainage piping. Interceptor is readily accessible for cleaning and inspection simply by removing the cover which sits flush with floor. The cradle supports the interceptor and has no provisions for waterproofing the interceptor in relation to the cradle and ceiling below.

Fig. No.	GPM Flow Rate	Grease Cap. Lbs.	Inlet and Outlet Size	Dimensions					C.O. Plug Size		Cradle Dimensions			
				Roughing Dimensions			Body Height	Body Width			J		K	L
			A	B	C	D	E	F	G	H	Min.	Max.	Length	Width
8304	4	8	2	7¼	3¼	16¾	11	9½	1½	–	5½	12½	20¾	13¼
8307	7	14	2	8½	3½	17	12	10¾	1½	–	5¾	12¾	22¾	14
8310	10	20	2	9½	3¾	21	13¼	12	1½	3¾	6	13	25	15½
8315	15	30	2	12	3½	25	15½	14¼	1½	3¾	5¾	12¾	28¼	18
8320	20	40	3	13¼	4	28¼	17¼	16½	2	4	6¼	13¼	32	19¼
8325	25	50	3	15¼	4½	30	19¾	17¾	2	4½	6½	13½	33¾	20¾
8335	35	70	*3	16¼	5	32½	21¼	19½	2	5	7¼	14¼	38¼	22¾
8350	50	100	*3	17½	6¾	35½	24¼	21¼	2	5½	9	16	41	24¾

* Available with 4" inlet and outlet when specified

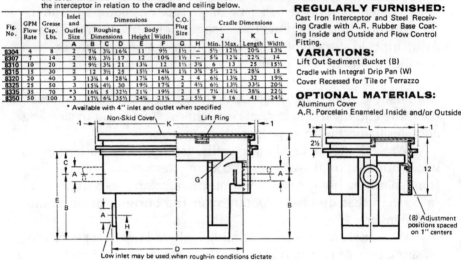

REGULARLY FURNISHED:
Cast Iron Interceptor and Steel Receiving Cradle with A.R. Rubber Base Coating Inside and Outside and Flow Control Fitting.

VARIATIONS:
Lift Out Sediment Bucket (B)
Cradle with Integral Drip Pan (W)
Cover Recessed for Tile or Terrazzo

OPTIONAL MATERIALS:
Aluminum Cover
A.R. Porcelain Enameled Inside and/or Outside

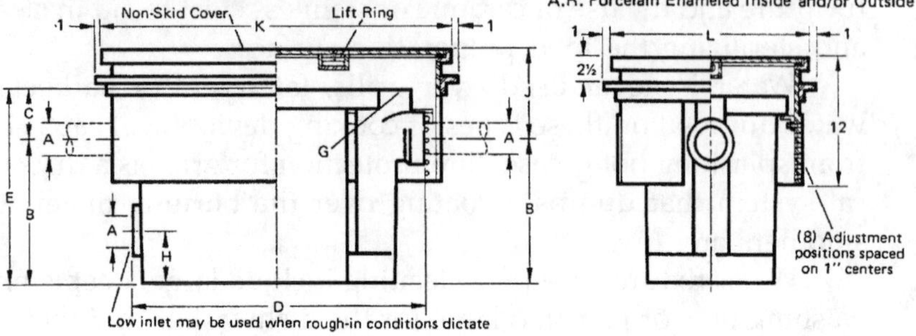
Non-Skid Cover
K
Lift Ring
1
1
C
A
E
B
G
A
B
A
H
D
Low inlet may be used when rough-in conditions dictate
L
2½
12
(8) Adjustment positions spaced on 1" centers

Figure 14-3. A grease trap. Courtesy: J.R. Smith Co., 1995.

Other types of water management devices found in or near kitchens include a mop sink. A mop sink is a large square sink that mounts on the floor in a janitor closet. The sink has a drain and a plug. The mop can be rinsed easily in the mop sink without having to spill from the wet mop onto the floor (see Figure 9-4).

Construction Materials

The construction materials used in kitchens are designed to be compatible with the use. Usually, because of the large amounts of heating elements and flame sources, construction materials in kitchens are non-flammable. Non-flammable materials include stainless steel, porcelain, ceramic tile and clay tiles. Wood is sometimes used for cutting boards and counters because of its non-slip quality while cutting and food preparation activities take place. The wood is left unfinished, since paints and stains should not come into contact with foods. Occasionally, plastics such as Formica or vinyl tile are used but do not provide the service life of metals and tile. The floor is clay tile with a non-slip finish, but vinyl and Formica floor tiles can also be used. Walls are block, tile or sheetrock. Sheetrock walls can be punctured and if so, punctured sheetrock walls become a source of potential disease and can result in health department fines.

The ceiling should be solid with surface mount light fixtures. T-grid or drop ceilings should not be used in kitchen areas as dust from this type of ceiling can affect the food preparation. Recessed fixtures will result in dust that can settle on food products when the bulbs in the fixtures are changed. In all, devices should be selected that are smooth and water resistant and that can be washed with wet soapy solutions.

Ventilation

As with bathrooms, kitchens should be well-ventilated to carry away smoke and odors from cooking and humidity from dishwashing, cleaning and washing down. Grills should be ventilated as well to carry away cooking fumes.

Mechanical Rooms

A mechanical room in a facility has limited access, usually only by staff and sometimes only special persons on the staff. As such, mechanical rooms are kept locked and are keyed only to those staff who need to have access to the rooms. Mechanical rooms will contain water treatment equipment (Chapter 11); heating, ventilating and air conditioning equipment; solar or hot water heating equipment (Chapter 12) boilers; hot water tanks; pumps; air compressors; and/or fire protection equipment.

A mechanical room, since it is not used by the public, needs not be architecturally fancy. Usually, a simple concrete floor, unfinished walls of block or sheet rock, open joist or beam ceilings and light fixtures to illuminate any maintenance activity are all that are provided. A mechanical room will have floor drains and some mechanical rooms will have dikes on the floor around the various pieces of equipment to keep a spill from one item from going onto another item. The floor, usually of plain concrete, will be sloped to the floor drains to aid in speeding draining. And depending upon the equipment in the room, a hose bib and hose will be provided to wash down and rinse equipment that has been serviced.

Mechanical rooms sometimes will be insulated to prevent noise from the mechanical equipment from spreading throughout the building. Insulation can be sprayed, batts or loose. However, if the insulation is not compatible with water used in the room, it should be covered to prevent moisture from water and wash down operations from saturating the insulation.

Swimming and Bathing Pools

Perhaps nowhere in the world does water and a facility come together more beautifully and functionally than at a large outdoor pool or combination swimming and bathing facility (see Figure 14-4). Here, the combination of cool sparkling water ramps, walk-ways, and grassy areas and people of all sizes, ages, and colors combines for pleasure, peace, comfort, exercise and recreation.

The plumbing and piping details for pools are discussed in Chapters 7, 8, and 9, and water chemistry and treatment are covered in Chapters 4 and 11. In this segment, the details of water/people interface are discussed.

Figure 14-4. A large outdoor pool complex, a center of fun, relaxation and recreation.

Fence

A swimming pool is designed to be attractive. The plaster, vinyl or tile pool lining is a light blue or green aquamarine color that has been carefully chosen and selected to provide an image of safe, warm bathing. This design is intended to entice the user into the water for pleasure, health and relaxation. This expectation will be realized for adults.

Every summer, it seems, however, a small child inadvertently gets into somebody's backyard pool and drowns. After many lawsuits, a legal precedent has determined the presence of a pool as an "attractive nuisance." This means that a normal person is enticed into the pool by its presence. Because of the attractive nuisance interpretation, the codes require that a barrier be erected about the pool to protect somebody idly passing by from being attracted into the pool to their own injury.

Most codes define the that a pool must be fenced. The fence must be such that a ball six inches in diameter cannot pass through the fence at any point. City or Public health officials will test the fence with a child's ball, pressing the ball at the larger fence openings. Most of the problems occur at corners or near the bottom where a small child could slip under the fence. The other problem is openings at the top of the fence provided the child is old enough to climb it. A public pool fence should be at least 7 ft. tall.

An unknowing owner might decide to construct a solid fence, where somebody outside the pool could not even see into the area. However, this usually works to the owner's disadvantage since it prevents the owner himself from looking into the pool from outside his property to the glee of teenagers and the dismay of the owner. The other advantage of an open weave fence is that security personnel and the local police can determine from quick observation if any unauthorized use of the pool is occurring.

In addition, the idea of having large openings allows the users inside to pass objects through the fence to outside family members and vice versa. Objects to be passed might include keys, money, towels, sunglasses, tanning oil, etc.

A pool fence should be well back from the water's edge to allow room to walk around the pool without interference. For a public pool, this is a requirement since a rescue could become necessary from anywhere near the pool's perimeter.

Walkways

A bond beam is constructed around the perimeter of the pool at the surface. The bond beam structurally holds the top wall in shape. Most concrete pools have a 6-inch wide bond beam. Walkways around the pool tie into this bond beam, usually with a construction joint.

Pool walkways can be concrete or clay tile, and some will have wood decking. Wood decking is acceptable but has to be replaced every few years to keep its smooth finish. With increased wood prices against a more or less steady concrete price, concrete has become the preferred walkway material near pools.

The finish of concrete can be a rough broom finish, smooth or stones placed in the top layer that can be rinsed, giving a rough cobble finish. If this latter method is chosen, make sure the stones are river-run or rounded. Some concrete plants use crushed stone which has sharp edges and these sharp edges will not work around a pool where the bathers walk in bare feet. If any walks are sloped, the slope should be less than about one inch in 10 feet to prevent patrons from slipping and falling on a wet deck. Likewise, any trenches or places in the deck that have been recessed should have a cover to keep the deck smooth, since one misplaced foot can result in an injury fall.

Access To The Water

A ramp, steps, a ladder, climbing out over the side, diving in, diving in from the side and jumping in from the side are all ways that bathers will get into and out of the water. Where the depth is shallow, signs along the deck should state: NO DIVING.

Stairs placed at the shallow end can be used by elderly and young patrons to enter and exit the pool. Stairs should have a rail to prevent slipping (see Figure 14-5).

For large pools, a ramp can provide wheelchair and therapy access. The ramp should have a non-skid surface to prevent falls and a handrail should run along the entire length to aid patrons and prevent slipping.

For the deep end of the pool, divers can use a board. The board should extend out over the water far enough to prevent the diver from coming back into to edge or end of the pool. For various diving heights, depths have been recommended by the United

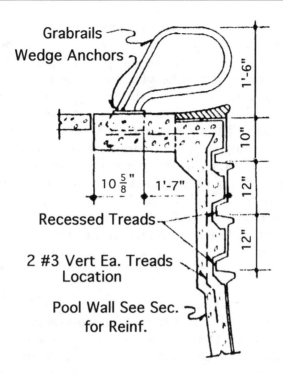

Figure 14-5. A typical detail of a handrail for a swimming pool.

States Olympic Diving Committee.

To exit a pool at the deep end, a ladder is provided. A ladder can be attached to the side of the pool, or the steps can be recessed into the pool wall with a handrail overlooking the side. The recessed ladder allows an extra lane when the pool is used for competitive swimming.

Finally, the side of the pool should make it easy for the patrons to climb out should they wish to. Jumping in from the side is perfectly acceptable, provided the water is deep enough.

Sunbathing and Shade

One of the pleasing aspects of a swimming pool is the opportunity for sunbathing. The combination of sunbathing and swimming are fun for all patrons. Areas for sunbathing can include the pool deck, a large grassy area or a specially designated sunbathing area. Lounge chairs, benches, wood decks or concrete decks can all be used. The sunbathing area should be back from the pool a reasonable distance so bathers do not get splashed. Since the shallow wading portion of the pool is heavily frequented by small children and groups of young children, the sunbathing area is often located at the other end of the pool. Recently, there has been a trend away from sunbathing and toward a more relaxed approach sitting in shade while watching the pool. Shaded observation areas within the fenced complex would be for parents to observe children and for those who do not plan to swim, but who need to be inside the complex as a result of a close relationship with bathers. Shaded areas can be permanent, semi-permanent or portable. Usually, an open pavilion is preferred. Contrary to the sunbathing area, the shaded area works best near the wading area of the pool.

Depths

Several functions take place within a pool including diving (platform, board), wading, floating, competition speed swimming, water ballet, water exercise classes, swimming lessons (beginning, intermediate and advanced), lap training, water team sports such as water polo, and in some pools kayak and canoe training and scuba diving.

This multiple use activity requires special allotments of pool space and even times of day when varying use is allowed.

Depending upon size and function, swimming pools will have wading/shallow areas, mid-depth areas for lap swimming and competition, and deep areas for diving. Floats and ropes are used to separate the areas. As with all pools, the floor should be clean, free of objects and obstructions and smooth. Wading areas are 3-4.5 ft. deep, mid-depth 4-8 ft. deep and deep pools 10-20 ft deep.

As the pool's depth increases, the volume of water in the pool increases geometrically. The increased volume leads to increases in pumping, purification and plumbing and piping costs.

Smaller hotel and residential pools usually eliminate the deep areas. Large public pools have all three types of functions and may even separate them into three separate pools for the three types of uses. Table 14-3 shows the type of pools functions and the associated estimated depths.

Staffing

Staffing needs for a large public pool include food vendors, access control personnel, plumbing/pipe mechanics, lifeguards, instructors and lawn maintenance personnel. In many pools, water chemistry tests are administered by the lifeguards but never during swimming hours.

Of course, the most significant staff at a pool are the lifeguards. Depending upon the size of the pool and of the complex, there may be one, two, three or more guards on duty. For private pools, management does not provide lifeguard services.

Table 14-3. Pool uses and recommended depths for each. Source: Utah State Department of Public Health, Regulations for *The Design, Construction and Operation of Public Swimming Pools*.

Wading	Exercise	Slide	Spa	Diving Height of Dive		
				0-3'7"	3'7"-9'10"	>10'
0-2'	5'	3'	4'	8'6"	10'	13'

Where a lifeguard is not provided, management should notify patrons of their risk and have publicly posted policies for safe use to be read by all patrons.

Management should recognize the function of lifeguards is to protect bathers. Giving lifeguards ancillary duties such as water chemistry tests, access management duties or other non-water-attentive duties increases risk that the guard will not see the occasional swimmer in trouble. Lifeguards should be rotated regularly and every effort should be made to support the lifeguard in providing safety at the pool.

Because of the risk of liability at swimming pools, many pools are owned and managed by the community. That is, they are owned by the city, county or township. This public ownership spreads the liability of the management and owners of the pool. On the other hand, hotel/motel pools are treated as private pools.

The services are provided for hotel patrons only, which limits the number of pool users at any time and reduces the risk. Municipal codes provide a guide to the approximate number of bathers per square foot of area. Managers should take steps to see that this number is not exceeded since both physical security/safety and water chemistry calculations have been based on this bather load number.

Pool Covers

Lifeguards should also cover the pool when not in use, since covering holds in the heat and prevents dust, grit, debris and trash from blowing into the pool when it is not being used. From a cost standpoint, covering a pool is economical since it reduces heating costs. Care should be taken when covering a large pool and horseplay among the staff kept to a minimum as a staff member who falls in, underneath a pool cover, cannot get out.

Indoor Pools

Most of the elements that apply to outdoor pools will apply to indoor pools as well. Access control is simplified, while shade and sunbathing issues do not apply. Consideration for walkways is paramount. The facility manager should recognize that as the floor space increases, the size of the indoor pool enclosure increases exponentially (by a factor of 3) and that costs mount accordingly. As a result many indoor pools, have large doors that open to the outside for sunbathing and other deck activities.

The advantage of indoor pools is that they can be used year round and benefits accrue from year round use. Exercise clubs, health spas, hotels and motels benefit from indoor pools. Indoor pools are smaller than outdoor pools with limited diving and competition lanes.

Indoor Pool Ventilation

The use of chlorine, iodine or bromine to sterilize swimming pool water plus the evaporation of the water itself creates a corrosive atmosphere inside enclosures of indoor pools. As a result, the air conditioning/heating of indoor pools has become a special field of indoor air quality designers and consultants. Heated pools will create extreme humidity which must either be exhausted out of the building directly or removed when recycled back through the ventilation system. The fans should have excess capacity to move large amounts of air and heating provided to make up for heat loss when room air is removed.

Water from warm humid air condenses on cold outside windows, defeating the purpose of having windows for outside visibility in the first place. Hard water chemicals or the salt residuals from softening will remain after the condensed water has evaporated leaving a chalk white film. All these problems can be eliminated with good design. However, if they have not been addressed in initial design, then maintenance staff will have to spend a lot of hours washing windows, wiping down surfaces, scrubbing walls and keeping the facility in model condition. As with the discussion on bathrooms and kitchens, indoor pool areas should be finished with water-resistant materials such as vinyl or ceramic tiles, block or brick and finished with water-resistant enamel or epoxy paints.

Aesthetic Accessories

Water parks, pools and indoor swimming complexes will benefit greatly from a few large potted plants, planters and other living flora. In addition, some vertical sculpture, modern art, and deck furniture will break up some of the flat horizontal lines prevalent at an indoor or outdoor pool.

Spas

Similar to the swimming pool is the spa. A spa or jetted tub is much smaller than a pool and is used by patrons for relaxing, health enhancement or recreation. Commercial fiberglass spas come preplumbed with all of the necessary equipment. Most of these commercial units are placed on a concrete or wood deck, wired in, filled with water and turned on. In general, these spas are for residential use and a facility manager should be careful about using one for pubic use. Public spas have additional plumbing requirements from spas designed for residential use. The public spa requires two pumps to recirculate the water—one for the jetting action in the tub and a separate, smaller one for recirculating the water constantly.

For both the swimming pool and the spa, pumps recirculate

the water, straining out debris while filtering and adding chemicals for treatment.

Heavy use of a spa or a pool can result in water chemistry treatment problems. Patrons perspire in the water. Chlorine treatment breaks down this sweat into ammonia. The ammonia in the water frees the chlorine, accounting for the chlorine smell at the pool. In addition, ammonia clouds the water and the water chemistry treatment equipment discussed in Chapter 11 is needed to purify the pool. Finally, chlorine test kits should be used to test the chlorine content and the water's pH. Pool water that is overtreated can be highly acidic which burns the eyes and has been known to damage teeth and skin.

Fountains

A fountain is sometimes installed at a facility for aesthetic purposes. The ancient Greeks discovered that the sound of quietly moving water was soothing and relaxing. Doctors and healthcare clinics have been able to capitalize on this soothing sound, and so have malls and the lobbies of some fine hotels and large office buildings. For a fountain, pumps, pipes and drains are used to recirculate water through a pool. Usually, the fountain-type pool uses plants and rocks to increase the "natural" setting. Large fountains are finished with cut stone or tastefully sculpted concrete.

Like pools and spas, most fountains have filters and water chemistry equipment to keep the water clear and clean.

One of the problems with fountains has been the tendency for vandals to discolor the fountains with food coloring or to add detergent, making large amounts of foam. Several spa and pool chemical companies sell a defoaming agent that reduces the suds that result from detergent. Unfortunately, there is not much that can be done for colored water. Usually, the pool is drained and refilled with clear water.

Like swimming pools, both spas and fountains can be constructed with perimeter tiles, lights, plants, and occasionally deck chairs.

Clinics

Another area where water and patrons come together that has a lot of visibility and requires unique design and management efforts are medical health facilities.

These doctor's office complexes include water fixtures with unique applications. Small community health clinics have taken the load off of some doctor's offices and reduced patron costs. Another facility that falls into this category would be dentistry offices.

Water management in clinics include providing the correct number of fixtures in the proper places and in maintaining the correct types of water supply and wastewater piping.

Both the emergency clinic and the doctor's office will have restrooms for staff and patrons and these are no different than restrooms in other facilities. Usually, however, the restrooms will be small and more personal since the persons using them will either be a member of the doctor's medical staff or the ill person and/or his immediate family.

The patient rooms in clinics will have a sink and wash basin within the patient room. This is to allow the medical person to wash immediately after treatment and prevent the spread of any disease to the next patient. Medical personnel have shown a preference for the use of foot valves for water flow and a long gooseneck faucet to make it easy to place the hands under the flow without touching sinks or the faucets (see Figure 14-6).

The patient room can have a soap and towel dispenser as well. The walls are typically sheet rock and the floor vinyl tile

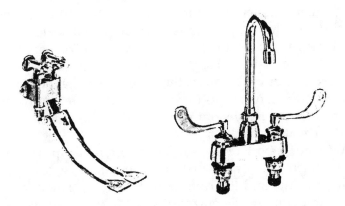

Figure 14-6. Typical medical Clinic foot valve and sink with gooseneck lavatory faucet.

which helps keep costs down when compared to ceramic tile and block. In addition, the clinic will have a central work station for the nurses to prepare medications, charts and syringes, as well as clean and sterilize the medical instruments and devices. Again, a sink and foot valve can be utilized for this purpose.

Unique maintenance of these fixtures is not necessary, nor is the construction. The walls, pipe, plumbing and vents are the same as for routine construction.

However, medical wastes such as blood, bile, etc. are normally captured with towels, bandages and cotton swabs. These are not rinsed into the wastewater system but disposed of in sealed bags. Since some of this waste is potentially hazardous, the container are carefully marked. Records of hazardous waste must be kept as discussed briefly in Chapter 3.

Laboratories

Laboratories are used to perform chemical and scientific tests. Laboratories use water for washing and sterilizing glassware and instruments, as well as for mixing, dilution and test preparation. Depending upon the functions of the laboratory, several stock chemicals will be used. Some chemicals are poisons, some flammable, while many are simply inert. Staff working in laboratories include chemists and other highly trained technical personnel.

Laboratory staff should understand the chemistry of the incoming water to determine if it needs to be treated before using it in mixing/washing operations. For many laboratories, small point-of-use reverse osmosis units or demineralizers are used (see Chapter 11). For large labs, the water purification equipment is centralized. For any laboratory, the facility manager needs to work closely with the lab staff to make sure their water needs are met since much of the results of the lab are dependent upon the quality of their water supply.

Water for laboratories will be of different types depending upon its functions. For example, the lab could use soft water for washing, rinsing, and the restrooms. However, the lab may need to use demineralized/deionized water for rinsing glassware and preparing analysis chemicals.

Therefore, a laboratory will have a long row of fixtures for various tests which include water, demineralized water and often several types of laboratory gases. All of these will be piped with the standard plumbing pipe and fixtures. In addition, some laboratories will sometimes have plumbing systems of double walled pipe with an inner pipe to carry the hazardous materials and the outer pipe to act as a barrier in case the inner pipe leaks (Chapter 7).

Double-walled Pipe

As reviewed in Chapter 7 double-walled pipe can be installed for laboratories or clinics where hazardous materials are generated. Double-walled pipe is expensive, although it can be installed in many areas. The joints, fittings and couplings are available in all types of materials. The facility manager's biggest difficulty in using double-walled pipe in a laboratory will be in finding pipe of satisfactory material that does not degrade from the various chemicals that could be used.

Double-walled pipe should have a sensor between the inner and outer pipes to indicate if a leak has occurred. The sensor sends a signal to the lab manager or to the facility manager notifying them of a potential leak that needs to be repaired. Selection of the leak detection sensor should be done carefully since a number of false alarms will cause all reliability in the system to be lost.

Many labs, upon careful evaluation, have found that the amount of hazardous waste can be minimized by careful water and wastewater management techniques, and it is possible the laboratory can live without the complex expensive double-walled system entirely.

In laboratories, rooms where chemicals are stored *should not have floor drains* since a spill of a hazardous chemical could corrode the waste water piping or poison the wastewater at the treatment plant. A spill kit—including a decontamination solution, rubber boots, gloves, rags and other absorbent material—should be kept in the room.

Chapter 15
Project Management

*P*roject management begins with good design, but good construction does not necessarily mean a good final product. It is entirely possible to do a good detailed design, write a great contract, hire an excellent contractor, finish the project on time and under budget, only to abandon the project because it does not meet the facilities' needs when complete. Good design is the result of good planning.

PLANNING

Planning is often the key to successful management in any facility. Unfortunately, there seems to be a severe shortfall in intelligent planning these days which leads to great frustration on the part of facility managers, staff and other professionals. Facility managers can also fall into the trap of continuous planning and not drawing planning to closure and proceeding with the next step of detailed design.

Planning At The Start
Good planning begins with goals. Successful facilities have a mission statement boldly and publicly posted for everyone to see. The goal keeps everyone focused on the mission and helps to recognize when the goal is accomplished. A simple goal for a facility water manager could be: "Reduce costs of water maintenance by 18 percent," while another could be, "Reduce the number of customer complaints to zero." Conversely, undefined goals leave everyone without focus and intent. "Our goal: provide water to meet the needs of the facility for the upcoming programs." What programs? Goals like these are for managers to measure their own success, not the staff's.

239

Goals should be realistic and feasible. Without everyone buying into the goal, it is not achievable.

With goals posted, sub-elements in the organization can further define individual goals and/or subgoals. Planning to accomplish the goals begins.

Planning should start broadly to allow for measurement of present conditions and options for attainment of the goal. Some decision-makers make snap planning decisions based upon their goals, others are intuitive while still others are careful researchers. A good leader will allow his staff to participate in the planning, if there is time.

Costs of Planning

Consultants will bid the costs of planning on an hourly rate, since planning can be nebulous and can lead to some unexpected and surprising results. The facility manager should be careful not to let costs dictate his planning. All too often, planning is driven by, "Let's see, we have about $25,000 in the budget, so lets come up with a project for that amount." Obviously this type of planning and budgeting process is fraught with potential for error. The original budget is not based upon facility needs but upon available funds. The planning process, therefore, is already limited because the goals of the project are not defined.

Given planning that results from budget allocations, the results of the project will not even be measured.

Long-range planning should be performed with an in-house staff because these people are more familiar with the facility and with the myriad details, markets and services provided. For a facility to pay an outside consultant to become familiar with these elements proves costly and the benefit of having them learn the details is lost after the plan is completed. Planning is an ongoing process.

Typical planning costs are shown in Table 15-1.

Economics do play an important role in the long-range planning process. Up until the late 1960s, for example, long-term growth had been steady for the previous 25 years and it was agreed to plan projects based upon growth rates of 2-6 percent. Rates of above nine percent were unheard of. Then in the early 1970s, rates and growth dramatically increased, peaking in the

Table 15-1. Project engineering and planning costs.

Long-range planning	$60-$125 per hour
Short-term planning	$60-$125 per hour
Feasibility study	3-5% of construction costs
Design drawings and specifications	12-20% of construction costs
Construction	Depends upon size of project. Plumbing for a typical building costs $6-$8/sq.ft. of the building.
Start-up	Depends upon the complexity of the project. For water systems, it is usually provided with the construction. Start-up of equipment such as water filters or softeners is included with the price of the unit.

early 1980s at the incredible rate of 16 percent. For many facilities, the planning cycle was driven into shorter durations. The quicker something could be done, the sooner the payback could begin. Long-term payback was ignored. Seven years later, the results of these short-term projects peaked and systems and facilities began to decline at a drastic rate. It is not uncommon to see building water systems built in the 1930s outlast systems built in the late 1960s. Many water systems built in the 1970s are nearing the end of their economic life.

There are still some very old buildings, of course. Notre Dame the famous French cathedral, was built in 1163 AD and is still used today. For its time, the most modern, ultimate construction techniques were used, and with the church funding it, the construction budget was more or less unlimited. Doubtful the facility manager will have such good fortune today!

Once the facility manager determines the life of the project or the duration of the planning cycle, the next step, which is factoring the economics, begins. The two major economic components to be considered are cost and benefits.

The Three-Stage Planning Process

Planning is essentially a three-stage process stemming from goals and moving to detailed design.

Stage 1

Initial planning is based upon assessment of the alternatives. Brainstorming, conceptualizing and defining scope are techniques used to come up with alternatives to meet the defined project need(s). This initial assessment is used to list all of the alternatives that meet some or possibly all of the needs.

Using this list, planners can evaluate the advantages and disadvantages of the alternative. A rough estimate of the benefits and costs is tallied. Benefits are compared to costs to highlight the more attractive alternatives. It is not so important that the costs or benefits be accurate at this stage, just that the tools used to do the estimates are the same to make a truly effective and useful comparison. Detailed work follows on the next step.

Stage 2

From the list of all alternatives, the most attractive two or three are selected. From these, more detailed estimates are made. The first stage was an attempt to separate the more attractive alternates from the least attractive. In the second stage, costs are estimated accurately enough to request funding. At this stage, although there still is not enough detail to proceed with construction, there should be enough detail to accurately determine the costs. Estimates at this stage should be within about 20 percent of the final costs. Factors for unknowns should be added to the estimates.

Stage 3

The final stage is where the detailed design is prepared. Several details are resolved here. The details should include the economic analysis of pumping versus pipe size (see Chapter 10). Cost of power should be checked and, if possible, future rates of power costs should be factored into the analysis. The final design would also analyze ongoing operating costs under the new system if possible.

Alternatives of pipe and tank insulation are made for a hot water system at this stage, with costs of insulation weighed against the costs of energy. Finally, costs of the alternative pipes sizes are made. Costs of seismic bracing should be included since some types of pipes require more bracing than other types. Initial pumps and tanks can be sized and cycle times and durations evaluated. From these times and cycles, costs of pumping can be weighed against the cost of tank sizes.

Assessing Benefits

Facility managers usually have to look to other sources to determine the benefits of a given project.

Benefits can be tangible—that is, their results can be directly measured in value of product sold or value of costs saved. This type of estimate should include competition prices, rate studies, marketing information and the like.

Benefits can also be intangible. Intangible benefits are those gray, undefined benefits difficult to precisely measure, such as the value of productivity among the workforce from drinking water of consistent high quality.

Intangible benefits and can become emotional issues in some cases. One example of an intangible benefit would be a "wild and free river." Obviously, there is a value to such a river but who can put a price on it? The facility manager's benefit analysis, therefore, should identify intangible benefits, even though a price cannot be directly assigned to it.

Assessing Costs

Fortunately, a dollar value for costs are much more easily determined than for benefits. The chapter on piping methods, and the chapter on equipment items both provide references for determination of costs. Elements of the project are broken down into units of labor and materials. The units are attached to unit prices and multiplied for a total cost. For planning, it is not necessary to determine the actual precise costs since the initial objective is to find the best alternative from hopefully several options.

On the cost side of project evaluation, there is risk. Estimates always contain some degree of risk. Risk can be minimized via field investigations and by checking and cross-checking the assumptions, calculations and raw data. No estimate can actually eliminate the risk, but if risks are known, decisions can be made that produce more accurate results.

One excellent source of cost data for facility managers is *Cost and Price Data* prepared by the R.S. Means Company. Means cost guides are prepared from public jobs bid throughout the United States. The R.S. Means guide also provides labor productivity data. For example, R.S. Means indicates that a contractor can place 350 linear feet of six-inch PVC pipe in one day. It also indicates that this work takes a crew of three men.

The planners assign priority to the projects based upon the benefits (both tangible and intangible) and the costs (including risks). It may be apparent that a large benefit can be attained at a small cost, or an alternate plan will attain a small benefit from a large cost.

Ike's Approach To Cost/Benefit

During his second term as president (1956-1960), Dwight D. Eisenhower directed that western water projects seek to attain a benefit-to-cost ratio of 1:1. That is, the benefits for the projects should be as close to one as possible. If a project was beneficial, those elements of the plan that were less desirable should be added to keep it from "making a profit." Ike's intent was to keep government projects from competing with private money-making ventures and to assure that the Government's investment was paid back, but just barely.

Refining The Plan

Once the attractive options are known, planners can concentrate on the best plans. In addition, plans can be refined, and attractive ideas from one or two low-benefit projects can be included with more attractive ones. The facility manager can continue to use the analysis for future planning and can keep the plan updated.

Other Planning Approaches

There are other methods used in planning besides a pure benefits/cost analysis. These included a weighted criteria ranking system and the "murder board" approach.

Weighted Criteria Ranking

Another method consists of establishing weighted criteria for each of the planning elements. A point system is then established based upon the weighted criteria and the options with the most points become the favored alternates.

Priorities such as cleaning up the water in one building is given a point score of 6, for example, with the highest score being 10. Replacing the water softener might be given a point score of 3.

The goals are also given a value that becomes a multiplier to the point scores. Cleaning water up might be given a multiplier of 2, while for softening the multiplier might be 1.

The combination of the goals multiplied by the point score for the tasks gives a number value assigned to the objectives.

The disadvantage with this type of planning is that the final result is usually one that none of the planners likes. Even though everyone agreed with the criteria and the weighting factors, the final project is not what anyone really wants. Hence, the project loses momentum and dies of its own accord similar to the efforts to work out the money split in the old movie It's a Mad, Mad World.

The weighted/ranking method can be successful, however, and if the facility manager has the opportunity to try this method, it is recommended if nothing else than as a learning tool. It helps define the critical issues facing the facility.

Murder Board Approach

Another method of planning has been called the "murder board." The author is not particularly fond of this planning method, but it seems popular with top management decision-makers and so it is presented here for discussion.

The murder board is a team or group assembled to prioritize and rank proposed projects. In this approach, the facility manager or his staff will go to a regional conference or annual budget meeting. In the murder board, the manager submits his proposals to the board along with the proposals from facility managers from other locations. The tactic is for the facility managers to fight it out over the budget, in an attempt to get projects funded for their facility.

The advantage of the murder board is that the project champions are instantly and clearly defined.

Unfortunately, however, the murder board depends upon the dynamic personalities of the various facility managers or board members. Some of the rational thinking necessary to good planning gets consumed in the emotion of the board hearings. The other advantage of the murder board is that everyone knows going in there will not be enough funds to go around.

The Finish Line
Once the most attractive plan or plans has been identified, the facility manager can proceed with final planning. Final planning should determine more realistic projections of benefits and of costs. Decisions about materials of construction should be made along with a schedules for implementation. Costs of construction should be estimated and funding for the project should be located.

DESIGN

Depending upon the facility manager's staff and access to consultants, a design is prepared from the final plan. Designs can be made in conjunction with contractors or consultants who may have better understanding of codes and construction materials. Often a contractor is brought in at the final stages of the planning to help out with estimates and schedules.

The facility should be careful not to commit to hiring any contractors at this time. Many times a facility manager becomes acquainted with one contractor, only to find that other contractors claim that fair and open competition was not sought because the facility allowed itself to become too closely aligned with one contractor too early in the process. Even if there is absolutely no truth to these allegations, if it looks funny to anyone, the facility manager is going to bear the brunt of the censure and charges.

Design Drawings
To explain the scope of work to contractors or facility construction staff, a final design is prepared. The final design consists of detailed drawings and specifications. Depending upon the

complexity, the detailed drawings can be large, carefully engineered drawings, small drawings or even hand sketches. As far as the contract is concerned, there do not have to be any drawings at all. The drawings are intended to represent what is to be installed, and where, for whoever does the work. For cities and large college campuses, these drawings become part of the master record and are used later by operations and maintenance to operate and troubleshoot the system. For underground work, the location of the pipes is tied by field survey to control points. These control points can be anything from a stake in the ground to a brass monument. Design drawings show the distances from the control points to the pipe and the direction the pipe is laid. After construction, it becomes a simple matter to locate the pipes.

If pipe is being placed in a city street, for example, the design drawings should identify any other pipes also in the street and the elevations so that the contractor/builder knows where the other pipes, power lines, telephone conduits, fire hydrants, sewer manholes and all the other physical features that will be in his way during construction of the pipe.

For work inside buildings, the drawings should show elevations—i.e., what floor, and where the pipe is to be located. Along with the drawings of where the pipes are going to be placed, the drawings should provide details that show all the other information the contractor will need to know in constructing the pipes. Location of fittings, tees, ells, cleanouts, pipe utility chases, hangers and details of pipe hangers should be shown. A good engineering staff will have some experience with construction and should be able to show the contractors the information needed to prepare a bid for the work. For facility managers who do not have an engineering or planning staff, a consulting-engineering firm is best equipped to handle the requirements of final design drawings.

Depending upon the desires of the facility manager, the plan drawings may be stamped by the designer. For pipe systems, drawings may have to be stamped by a registered civil or mechanical engineer because it may be required by local building codes.

Sometimes, the designer provides notes and specifications right on the drawings and eliminates the text specifications altogether if he desires.

Final Specifications

The final specifications include written requirements for the materials of the piping, sizes, valves and fittings. The text references applicable national codes and standards and spells out the requirements for testing and coordination between various trades. The text should also reference the drawings.

A typical specification will be organized as shown in Figure 15-1 which also shows other components of the construction contract. Depending upon the complexity of the project, a detailed specification can be from 25-1,200 pages long. The extra length is for more types of materials. Each part of the detailed specifications relates to one of the parts of the project. For example, there will be a chapter on pipe. The pipe chapter will reference industry standards, materials, delivery, handling, factory tests, manufacturer certifications, quality and related work. The next chapter, which may be electrical work, would specify industry standards, wire, materials, delivery, handling, factory tests, quality and related work. As the number different types of construction are specified, the length of the contract specifications continues to grow.

Most engineering firms use standard specifications that are tailored for the individual project. The American Institute of Architects and the Construction Specifications Institute both publish standard text specifications that are edited by the designer for the specific job. In addition, the U.S. Army Corps of Engineers, the U.S. Veterans Administration, and the U.S. Naval Facilities Command each publish their own standard specifications as well. The Government Standard Specifications are in the pubic domain and can be copied without paying any fees other than the costs for copying. The standard specifications can also be directly downloaded via computer modem although fees are charged for access. (See Chapter 19 for contact information for the American Institute of Architects and the Construction Specifications Institute.)

After the detailed specifications are written, the contract should add any preliminary chapters. These are sometimes called the standard clauses because they are standard to all of that facility's contracts. They are sometimes known as "boiler plate."

Standard clauses include such issues as contractor billing for payment, method of payment, hours of work, schedule, utility tie-ins, contractor use of restrooms, and heating and lighting during

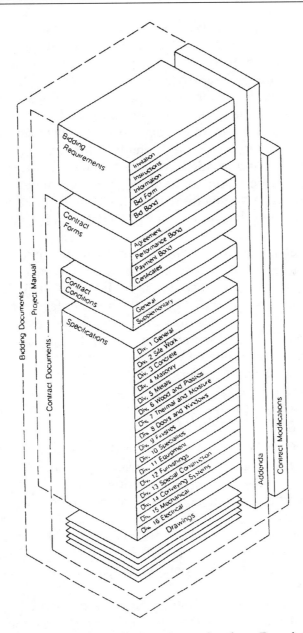

Figure 15-1. Construction contract organization. Reprinted from: The Construction Specifications Institute (CSI), Manual of Practice, (Figure FF/CD-4, FF Module, 1992 edition) with permission from CSI, 1996.

construction. Standard clauses also include wording about how the contract can be terminated, bonds, guaranty and insurance during construction. In addition, government contracts include references to the Federal Acquisition Regulations (FAR). The Federal Acquisition Regulations incorporate the many laws passed by the Congress. These include buying American Goods, Labor Rates from the U.S. Department of Labor, rules against bribes and kickbacks, contract termination for default or for convenience. In general, all of the nation's rules and laws that have been passed in the interest of fairness and competitive spirit are added to government contracts. On government contracts, there is often so much more paperwork that contractors add extra staff just to keep track of the documents.

One final word for facility managers about contract specifications and drawings: if there is a dispute between the contractor and the facility and the dispute cannot be resolved and the parties sue each other, courts have generally ruled in favor of the text of the specifications over the drawings. If the drawings say one thing and the specifications say another, the contractor, in effect, has a choice about which one he wants to do. Inspection resolves these issues early, during the work.

FINAL COST ESTIMATE

When the final design has been completed and detailed engineering drawing and specifications are complete, the cost of the project is estimated by the engineering firm or facility staff that has prepared the design. Some facilities require this estimate to be sealed pending the results of the bidding.

The work of the final estimate can proceed in parallel with the addition of the boiler plate if the project schedule is critical. For any job, the intent is to obtain an accurate estimate of the costs of the project.

Sometimes, however, in the thick of a disaster or on a project that is really rushed, the facility manager has to go with the best estimate he has. The best estimate is one prepared with the best information, but many water jobs are started, performed and finished on the basis of a firm handshake and a commitment to a vague dollar amount. When this happens, there is not much the facility manager can do, because there simply is not time to prepare all the designs and estimates. An emergency situation must be dealt with. The safest thing a facility manager can do in this case is have the contractor keep track of the actual labor hours and equipment used. Later, this information can be used for payment.

FAST-TRACK CONSTRUCTION

For a job where the project needs to be completed in a hurry, the construction team can build a project on what is called a fast-track schedule. Fast track is where the design of the element immediately precedes the work. For example, the pipe for a large pipe job is determined to be 36 inches of ductile iron. The contractor is told to go ahead and purchase the pipe, the details of where it will be placed is provided later. Next, the alignment is determined and surveying begins. In fact on a cross-country pipeline, it is possible to start digging trench for the pipe on one end when the exact route in the middle is not known. The advantage, of course, is the speed with which this type of project can be accomplished, the disadvantage being that decisions made early on cannot be changed later without a tremendous increase in cost. Recent construction trends do not favor this type of construction because the cost increases are often significant.

HIRING CONTRACTORS

Facility managers have a tremendous opportunity and latitude when it comes to hiring contractors. Basic contract law says that where there is an agreement between two parties where one agrees to pay the other for work, there is a contract.

Most disputes in contracts involve disputes between the scope of work and method or amount of payment. Written contracts try to reduce the opportunity for dispute. Contracts can be for services, for construction, for materials, for labor, for tests or for anything else that is legal.

Service Contracts

When services are needed the contract is called a service contract. For example, a facility manager may want to have a laboratory sample taken and a report written on the water quality on a monthly basis.

A letter agreement can be used whereby the facility manager requests letter proposals from competing labs and the one selected is sent a letter to which his quote is attached with instructions of when to begin sampling, where and how to bill and receive payment.

TYPES OF CONSTRUCTION CONTRACTS

More complex projects, including construction, can be contracted in various ways. Some types include unit price, fixed price, cost plus fixed fee, cost plus a percentage of cost, and cost plus award fee. Essentially, the facility agrees to pay the contractor the cost for work or services rendered plus some profit. The profit becomes the contractor's incentive to successfully complete the work. Complete details of the alternative successes and failures of these various types of contracting mechanisms are beyond the scope of this text—however, the facility manager can utilize references at the end of the book to learn more about the advantages and disadvantages of these alternative forms of contracts.

Fixed Price Contracts

For most projects, the facility can utilize the fixed-price method of contracting. The fixed-price method has been successfully used, it is reliable, and it has been tested many times in court by both facility managers and contractors. It is therefore one of the most recognized methods and should be the easiest to implement and use. The facility manager can hire an engineering firm to represent the facility in the bidding process or the facility manager can take delivery of the designs from the engineering company and continue with the project on his own.

For facility managers using public funds, it is in the best interest of the public for the project to be bid among all of the interested contractors. In this way, competition between the contractors assures the taxpayers who are funding the project that the best price has been attained. For private facility managers, a competitive bid is not necessary, provided the managers are confident that the contractor's bid and the work to be done will be satisfactory.

In the fixed price method, the facility manager gives the detailed design to the contractors and requests a bid or bids.

Working from the plans and specifications, the contractors calculate the materials they will need and the man hours of labor to get the job done. From these he prepares a bid. If the project is being competitively bid, the manager should request the bids be completed within a fixed time and that all contractor's have the same amount of time and the same information.

Occasionally, the contractors will point out a discrepancy in the plans that has to be clarified. This can result in a bid extension, depending upon the size of the discrepancy. Care should be taken not to give one contractor information without giving it to the others, or unfairness can be claimed.

Finally, if the project is competitively bid, the facility manager should insist that the contractors take no exceptions or their bids will be rejected. Under the fixed-price method, contractors realize that a mistake could cost them dearly, and as such will want to be careful their bid is correct. Uncertainties cause the contractors to want to hedge their risk and hence they attempt to add some type of exclusion or caveat to the quote. If the bid requires no exceptions, the facility manager can reject the bid, even though he may decide not to. This is a nice out for the facility.

Once the bids are opened and tallied, the lowest bidder is known. If the bid is acceptable to the facility, a contract is awarded or signed with the low bidder. On a large job, this sometimes can be exciting but often it is a rather dull affair, especially for large contractors who bid a lot of jobs.

For private facilities, it is not necessary to award the contract to the lowest bidder, but to the contractor that the facility decides has offered the best combination of price, schedule and resources. Facilities may want to check the contractor's work on other jobs and to talk with other facility managers who have utilized the services of that contractor.

FIELD WORK

Once the bids have been tallied and the facility has decided who the contractor will be, the facility manager and contractor should hold a preliminary meeting to discuss the work. Since all of the other bidders have been eliminated, the facility manager and the contractor can now finalize the details of construction.

A Word About Contractors

There are many types of contractors, some coarse and some refined. Occasionally, there is a disreputable contractor but these generally do not remain in business long. A common mistake that facilities make in dealing with contractors is that since the specifications are so detailed and so much money was spent on the engineering consultants to prepare the design packages, there is little give-and-take in the course of progressing with the work. For many years, contracting specialists have operated under the theory that if some matter is not addressed in the written word of the contract, then it does not exist. This narrow-mindedness ends up costing more money than it saves, but the problem stemmed from too many gentlemen's agreements in the field between the facility manager's staff and the contractor's staff.

Even with the construction specifications and plans and drawings, there is room for differences and for resolution of differences. For example, the weather plays an important role in any construction since rain, snow and other wet conditions affect the production of the crews, access to the work, and inspection of the work.

The preliminary meeting should address the contractor's initial plans for schedules, materials, parts, use of parking lots for crafts and labor, safety requirements, and the contractor's use of electricity, water and sewer systems. Many is the facility manager whose contractor shows up on the first day of the job and blows out the fuses in the facility manager's headquarters. In the dark and cold, the facility manager begins to wonder if remodeling was such a good idea in the first place.

In addition to utility services, the facility manager should discuss billing and payment at the preliminary meeting. The inexperienced manager soon realizes the contractor is in the process of purchasing materials, hiring craft and labor, ordering tools and equipment and numerous other details.

An example of where a facility manager can get into

trouble is with equipment. If the contractor decides to borrow the facility manager's equipment—a backhoe, let's say—and while in the process of digging with it, the hole caves in and a worker is injured, the contractor, to protect his own liability, could claim that there was something wrong with the backhoe. This makes the facility manager potentially liable for the contractor's mistake. Therefore, it is generally recommended that the facility not make any equipment loans to the contractor in order to remove this potential facility liability.

After the preliminary meeting, the contractor will begin work by bringing equipment and materials to the job. The facility manager should arrange to have the work inspected every day, or more often, depending upon the conduct of the work. The facility manager's best tool for inspection is a camera. Photographs of construction are influential and put the contractor on notice that his work is being recorded. Later, professional engineers, lawyers, contracting personnel and payment personnel can examine the photos to determine if there are any discrepancies. Even video cameras can be used to walk through the construction site and record the work as it is being completed.

The Construction Progress Curve

For a big job, the progress of the work follows a curve similar to the one shown in Figure 15-2. The "S" curve indicates that work and progress start slowly while holes are dug or demolition of old materials that are in the way are removed. Then when the preliminary work is done and tools, craftsmen, and materials are all on site work progresses more quickly until the bulk is completed. This is represented by the steep part of the curve. Then the work slows down while the final details of the project are being completed and testing and cleanup is done. Finally, the project is finished.

A facility manager can profit from knowing his involvement in the project occurs when the curve is changing shape (slope). In these areas of the progress curve, changes are occurring, and in this time frame the contractor and facility managers have to work closely together to keep the project under control.

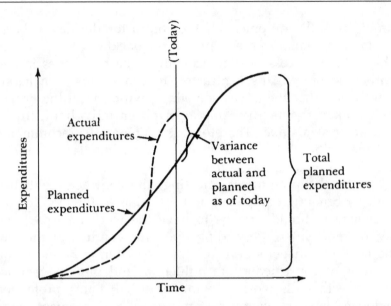

Figure 15-2. Reprinted from *Managing Projects in Organizations* by J. Davidson Frame, with permission of Jossey-Wales Co., 1991.

INSPECTION OF THE WORK

Inspection is the facility manager's way of assuring that the work is performed according to the design and that the final product meets the *facility requirements*. Note that the final product meets the facility requirements, not the specifications and bid documents. This is an important point that facility managers should keep in mind when a construction project is underway. There are a great many ways a project can go awry and the main purpose of inspection is to keep the project from getting out of control.

A facility inspector already knows a lot about the facility before the contractor begins work. The inspector can warn the contractor when utilities are at peak use so that utility outages can be scheduled for other times. He will know the locations of many utilities and can tell the contractor where they are. The inspector should have a list of phone numbers and be able to coordinate activities of the contractor with the facility manager's counterparts without having the facility manager handle all of the day-to-day contacts, issues, complaints and other stressors that can make a facility manager's day long indeed.

In addition, the inspector observes and watches the contractor's work. He can protect the facility from potential lawsuits from either the contractor's workmen or the contractor. The inspector makes sure the contractor uses the materials and products required. A classic example of this is in the installation of copper pipe. It is difficult to tell the difference between Type K and Type L copper pipe from a distance, although it is printed upon the long runs of straight pipe in ink. All the inspector has to do is look and make sure the pipe being used is the pipe specified in the job.

Unfortunately, many inspectors do not have the engineering technical training to be able to determine if one type of copper pipe is more suited for a particular job than another. Their training usually involves inspection to the specifications or to the codes and standards under which the work is being done. The facility manager needs to recognize that the inspector is protecting his perceived interest in the project and in the facility. The inspector may need to discuss details of the design with the engineers to make sure the contractor is using the correct materials.

If a facility manager allows the contractor superintendent to contact him personally and not through his inspector, then the inspector's position will be undermined. If the contractor meets with the facility manager without the inspector present, for example, the contractor can then tell the inspector that he has already discussed and resolved an issue with the facility manager. The inspector, not having been at the meeting, has to check with the manager to verify if this information is correct. If the facility manager is busy, in a meeting, or otherwise unavailable, the inspector either has to stop the work until the information can be confirmed by the facility manager, or allow the work to proceed with the risk being that the facility manager will later chastise the inspector for allowing substandard work. There is also the opportunity for the contractor to claim he misunderstood the facility manager, which is common.

The relationship between the facility and the contractor under contract law is an adverse one. That is, the contractor, who desires to profit from the work, will perform the minimum amount of services and construction according to the contract. (This turns out to be the case in fixed-price work; however, in cost-plus work, the contractor, who gets all his expenses paid, will go to great lengths to do extra work that he can bill for, which the inspectors should stop.) The contractor keeps the facility from gold plating and spending too lavishly on the project while the inspector keeps the contractor from taking short cuts, using substandard materials, untrained workmen or other practices which lead to poor work. The legal term often used is "at arms length," meaning that the contractor and the inspector must not become too friendly.

Inspectors by nature become suspicious and this author has never yet met one that has not seen some type of cheating done by one or more contractors. Their duty calls for them to be on the ground with the workmen every day. If the workmen crawls into a narrow confined hole to patch a leak, the inspector may have to crawl in also and verify the patch has been done correctly.

The facility manager must trust his own construction inspectors and if he finds he cannot, they should be replaced. Conversely, the facility manager should never completely trust the contractor, because there is too much at stake.

Finally, the inspector should keep a record or log of his activities, which should include the dates he inspected work, what was inspected, issues discussed with the contractor and the weather if it is an outside job. It is not necessary for these logs to be long, flowing prose or detailed descriptions of conversations with the contractor, but enough data should be recorded so that the day or events can be recreated later. The inspector should make sure the contractor keeps records on the drawings so that work that is later covered up is recorded. Finally, the inspector should take pictures of the work performed by the contractor.

The facility manager may wish to schedule a daily of weekly meeting with his inspector to remain current with the work. If there is not time for a meeting, the facility manager should direct that the logs of inspection and the photographs be provided to him routinely. It is not wise for the facility manager to take information from the inspector second-hand, such as through a secretary or a clerk, since relaying of messages can be confused. Ideal construction takes place where:

- The bid is low.

- The contractor did not leave out anything.

- The contractor is familiar with the facility and has worked there before.

- The contractor has done his homework.

- The contractor knows his work will be thoroughly inspected by competent inspectors.

- The contractor is paid promptly for the work after it is finished.

Unfortunately, many construction projects never make even half of these criteria!

PAYMENT OF CONTRACTORS

The other major contract provision in construction is the contractor's payment for work completed. First, a contractor on a project is buying materials, hiring labor, renting equipment, purchasing tools, etc. In addition to these, he will hire other companies to do work that he does not normally want to do or does not have the skill and expertise to do. On a swimming pool job, for example, the plaster lining may be done by another company. For a large sewer job, the contractor may hire a specialty excavation contractor to handle the digging and backfill, while he performs the critical tasks of laying the pipe and testing it himself.

All this type of work and material requires payment, and before the contractor can pay others for this work, he wants to receive payment from the facility. Usually, the contractor prepares a bill and requests payment. The bill may be for a percentage of the work, for all the materials ordered so far, for subcontracts entered, for quality testing, for surveying or any of the other tasks. In general, the facility manager should not pay the contractor until the work is done.

The author's personal policy is never to pay for pipe on a trench job until it has been placed in the ground. If the contractor has a yard full of pipe, there is no excuse to bill the facility for it. Even though the contractor bought a truckload of pipe to install in the facility, there is no guarantee that the pipe will not end up on some other hot job of the contractor's and he will order more for the facility manager's job.

Since most contractors have monthly billings from suppliers, contracts usually allow the contractor to bill every month. It is not necessary to bill every month and some jobs are billed by the contractor after the job is complete. This works especially well if the job is a small one.

Payment can be based upon a unit price method of work completed. For example, if part of the job calls for 1,000 ft. of pipe and 500 ft. have been installed, the contractor can bill for 50 percent of the agreed amount and the facility can choose to pay it. However, if this method is used, the facility manager should hold some back for testing, since the final tests require labor and materials as well.

Payment can be made in one lump sum at the end of the job. However, the contractor is likely to borrow the money to pay his subcontractors and suppliers. This cost of money is part of the job and the contractor often includes it in the contract price; the facility ends up paying these extra costs through the contract payments.

Payment can be based upon time with an even distribution over the time to perform the job. If the job takes six months the facility can pay 1/6 of the job each month.

Finally, contract payment can be based on a portion of the work completed as shown by the progress curve in Figure 12-2 earlier in this chapter. If the progress curve indicates that 25 percent of the work is complete, and this percentage is confirmed by field inspector reports, then payment of 25 percent of the job can be made. Use of the progress curve for payment and job tracking has many sophisticated uses for payment and performance as well.

As shown in the attached figure, the contractor's performance fee could be based upon how well he follows the curve. If he exceeds the curve by good performance, he might be offered a bonus, and if his progress lags behind the curve, perhaps additional funding should be retained as an incentive to accelerate.

SHOP INSPECTION

Perhaps pipe is being fabricated in a shop, or one of the equipment items such as a large water softener is being put together at a factory. In these cases, if the facility manager has elected to make a partial payment to the contractor for this type of work, the facility should arrange to have it inspected at the factory or shop. In addition, the inspector should verify the materials are slated for the facility manager's jobsite.

More than once, a contractor has been tempted by an urgent order from another project to ship the equipment to a different direction. On one project, the U.S. Government urgently needed a large pump a subcontractor was putting together for another job. It was shipped to the government job much to the facility manager's dismay. Later, the Government reimbursed the facility and the contractor for the delay. The Government need was that urgent.

This kind of incident is rare, but it can happen. The facility manager should exercise care to protect his funds and budget when making partial payments.

BONDS AND BONDING

Bonds are a way of assuring the facility manager that the contractor will complete his work within the required budget and scope. The bond is, in effect, insurance that the contractor's work will be done. There are several types of bonds, but not every type of bond applies to every construction project. The bond costs the contractor money and ultimately, this cost is passed along to the facility in the contractor's proposal.

Before any discussion on bonds, the facility manager should know that a recent trend is to eliminate bonds in construction projects since many bonding companies fail to take over the project when it comes time to do so. In fact, on one project the bonding company hired the original contractor to complete the work! If the facility manager is confident that the contractor is honest, reliable and has the resources to get the job done, it may be worthwhile to eliminate bonding altogether.

For those facility managers who require bonds, the three main types of bonds are bid bonds, payment bonds and performance bonds.

Bid Bonds

In a bid bond, the contractor guarantees that he will complete the project for his bid price. The bid bond then accompanies the bid and is the contractor's "deposit" that he will perform the work for the bid price, plus any changes, of course.

Bid bonds cost 5-10 percent of the bid price and are forfeited if the contractor fails to start work if his company is the low bidder. Bid bonds are often eliminated from the process, but the advantage of the bid bond is to keep the contractors from submitting courtesy or frivolous bids. It also encourages the contractor to keep from making mistakes since the bond can be held by the facility until mistake is proved and this ties up some capital the construction company could use on other projects.

Performance Bonds

A performance bond is a commitment by the bonding company that the contractor will perform the work of the contract. In contrast to a bid bond, which is a commitment to the bid, a performance bond is a commitment to the project or to the entire job. Usually, the performance bond is submitted after the bids and before the construction work starts. It is evidence that financial backing is present should the contractor fail in performance of the work.

Payment Bonds

A payment bond is a guarantee that the prime contractor will pay his subcontractors. On some contracts, the main (prime) contractor bills the facility but then, for some reason, does not pay his subcontractors with that money. Under law, since the subcontractor's labor and materials are in the facility, the subcontractor has the right to sue the facility for payments he has not received from the prime contractor. The fact that the facility has paid someone else does not relieve the facility of the obligation to the subcontractor. Therefore, the payment bond is the prime contractor's guarantee that he will pay the subcontractors. For most facilities that use bonds, this is the most important bond to have on the job.

A Final Word About Bonds

Facility managers should realize that bonds and insurance are generally handled by attorneys and lawyers. And this text does not pretend to give the facility manager any legal advice. In the event decisions are made to include bonds or that bonds are in question, it is wise to seek the advice of counsel in this area.

Finally, the bulk of the work is complete, the crews are released and testing can begin.

Chapter 16
Performance Testing

A *successful construction project will include testing. The facility manager wants to hold the construction contractor responsible for testing since the contractor is installing the pipe and equipment. In addition, the facility will often conduct its own tests after maintenance work has been completed. Field testing ensures that the installed system operates as designed and intended.*

FIELD TESTING

The facility manager wants to be assured that the finished water system product performs as it is intended. For piping systems, there are two types of tests—pressure and flow tests. For equipment, the facility manager wants to make sure the system is tested to make sure that it does what it is intended to do. If the facility has a new water softener, it needs to be tested to verify that it removes the hardness, and that it removes the hardness from the specified amount of gallons before it has to be regenerated. Chlorinators, pumps, valves all need tests to make sure they work. In short, water systems, like other building systems, must be commissioned.

Pressure Leakage Tests in Pipes

Pipelines that leak do not do anyone much good. Inside buildings, leaks will eventually be identified because the water will show up somewhere. Underground lines can leak, but the facility may never know it because the water drains underground to the water table. Leak tests make sure the system is free of leaks.

Usually, leak tests are a two-stage operation with a gross leak test performed to make sure all the equipment has been installed correctly and a design pressure test to make sure the pipes will hold design pressures.

To perform a leak test, pipelines are pumped up to test pressure with either air or water. The pressure is held for one, two or, for some systems, 24 hours. The pipe is then inspected for leaks and if no leaks are found, the test is accepted and the pipeline is complete.

Records of the test should be taken and signed by both the contractor and facility. The results of the tests are included with the final documents.

Pressure Test Methods and Safety

Pressure tests are most safely conducted with water. Air tests require the pipes to be pumped up using pressurized cylinders or air compressors and if the pipe fails during the test, expanding air can hurl pieces of the pipe in many directions. A water test does not cause this sudden expansion. On a large job, the contractor may do an initial leak test with air, looking for gross leaks in a large system, then go back and do a demonstration test for the facility manager with water. For a gross leakage test, the pipe is not pumped up to full pressure. Just a small portion of the test pressure is applied to confirm any gross leaks.

It is a good idea to use air for the gross leak test since gross leaks are fairly obvious. Most are the result of a valve not being closed, or a flange left out of place. On some pipe systems, the fittings may not have been tightened or perhaps one of the plumbers forgot to glue a joint together. If the gross test is done with water, this can cause a large mess that can be avoided by using air.

Once the contractor is satisfied that there are no major leaks, the pipe is then filled with water and pressurized to the design amount for the specified time. Often, a contractor is tempted to continue using the air pressure method since he has already set up the equipment. This decision should be up to the facility. The issue is whether to accept the risk of the air test, should it fail, versus the extra costs of setup for filling the pipes with water, pressure testing with water, and disposing of the water afterward. The decision depends upon the size of the pipe since it takes a small air compressor a long time to pump up a large pipe system.

A pressure test with air also requires the use of a soap

bubble solution to look for small leaks where leaking water would be obvious. The other parameter to consider in deciding whether to test with air or water is obtaining the necessary water, and where to waste it after the tests have been run. If the drains are completed, the drains can be used.

Finally, with both air and water tests, inspectors can sometimes become confused if it is a long test because of atmospheric pressure and temperature changes throughout the day. These ambient changes will affect the test. If the pipe is filled with warm water on a hot day and later that night the pressure is read again, the pressure will have decreased slightly because of the cooler temperature of the water and the water contracting as it cools.

The American National Standards Institute has specified an amount of allowed pressure change as a function of temperature and volume. This information is sometimes included with the specifications. However, it becomes important if the pressure changes over a long timed test as to whether the thermometers have been calibrated correctly, and whether the volume in the pipes has been correctly calculated. All this kind of information can become confusing during the test and should be spelled out beforehand. The results should be included with the test report.

Of course, if the test fails, the facility manager should insist the entire test be repeated. The contractor, while locating and repairing a leak in one spot, may cause a leak at the next joint while fixing the first leak. If the test is not repeated, the facility manager would not know the pipe was leaking at the new location.

Furthermore, the manager should insist that the test be repeated until the system *passes* the test. On one job, a contractor performed the required test, identified the leaks and said that to fix the leaks was extra work, since a test was specified but passing the test was not a contract requirement.

One last thing about pressure tests. They are intended to be nondestructive tests. This means the facility does not want the contractor to break the pipes while testing them. However, the system should be tested enough until there is confidence the pipe will withstand the operating conditions. A test that does not represent true operational conditions is of little value.

Flow Tests

After the pressure tests, the contractor may be directed to conduct a flow test. Not only is it the intent of the pipeline not to leak, but the pipes should deliver the correct flow to the right locations. An obstruction in the pipe can limit the flow which, of course, is not the intent of the pipe system. An excellent example of a flow test requirement is included in the National Fire Protection Association's Fire Sprinkler Codes.

Flushing Lines

The line should be flushed, then filled with clear water. The flow should be allowed to flow through the valves, or the pumps are turned on. Flushing the line is always a good idea since the author has seen rocks, gloves, welding rod, soda cans, sticks, weeds and a number of other types of debris come running out of a pipe while it is being flushed.

Setting Up The Flow Tests

Before flow tests can be conducted, a couple of items should be resolved. A decision should be made as to the disposal of water. If the water is treated and chlorinated, it may be necessary to reduce the free chlorine before disposing of the water since too much chlorine could affect fish or microorganisms in a nearby stream.

Since the flows are going to be measured during a flow test, the meters should be checked to make sure they will accurately record the correct amounts of flow. The tests should also be conducted with calibrated meters and the dates and records of the tests noted in the test reports. As always, a representative of the facility manager should witness the flow test.

Facility management staff should keep in mind that the test is conducted by the contractor and no action should be taken by the facility manager's staff during a test since a pipe failure resulting from action by the facility manager will result in extra payment to the contractor. Flow tests can sometimes be affected by the outside weather conditions, but not to the extent that pressure tests are affected. Since Chapter 10 presented a discussion of the relationship between flow and pressure, it will not be repeated here other than to say that the flow test is a measurement of system performance based upon theoretical calculations that are similar to the real flow conditions. If there are problems reaching the required flows, there is potentially an object in the pipe or air bubbles that are obstructing flows.

Failing The Test

If the system does not meet the required flows, the facility manager is faced with a classic dilemma faced by all facility managers involved in construction.

The installers will argue that the system was not designed correctly, and therefore the construction contractor cannot make the system perform as intended because the flaw is in the design, not in the construction. Therefore, the contractor has no obligation to the facility since the design was done by someone else who worked for the facility manager and hence, failing the test is not his fault.

The issue here is one of whether there is an object inside the pipe blocking the flow. First, the facility manager wants to consult with the designer. It may be that there is an error in the calculations or that the design assumptions were incorrect.

Keep in mind that even the most accurate of flow calculations are only within five percent of the true flow. Usually, the decision to select the "next larger size" or pipe diameters gives enough extra capacity so that this problem is solved in planning.

If the decision is reached that there is an object in the flow, the contractor is expected to locate and fix it. It is part of the job because the contractor has agreed to install a system that meets the flows. Use of an inspection camera may help to locate the problem (see Chapter 17).

Pump Tests

Pump tests can be conducted in conjunction with the overall flow test, or a loop can be installed simply for testing pump flows. The pump test should check the performance of the pump according to the curve data shown in Figure 18-4. Usually, a pump test seeks to verify the pump curve by measuring flows at three points and plotting the pump curve for the installation. The three points are connected and the curve is presented as the final pump curve for that system. The tested pump curve should be turned over to the facility and saved for future reference. If the facility has to be remodeled at some future date, the new designer can use the pump curve for design calculations.

Electric Motor Tests

In conjunction with the pump test, the electric motors are tested to confirm their performance in the installation.

For the larger pump motors, the power lines to the pumps are checked to make sure the insulation was not cut or nicked during installation. This is called a high potential ("high pot") test. For the high pot test, the motor is taken off the wires and the power lines to the pump motor are subjected to high voltages to determine if the wire insulation meets the design criteria.

Some facilities delete the high pot test since the wire insulation can be degraded by this type of test. For an existing facility, a high pot test carries some risk. The powerline could fail as a result of a shortage. If the line fails, it has to be replaced and the equipment is down while the power is being repaired. For a high pot test in an existing facility, the manager should have a backup plan in event the powerlines short out.

Next, the pump motor needs to be checked to make sure the coils are correctly wound and there has not been any damage in the motor windings. For this test, a meggering meter is used to measure the electrical resistance of the windings. Meggering is usually an excellent method to ensure the motor will perform as designed.

Electrical Safety

Water and electricity do not mix, so the facility manager must exercise caution when conducting power tests or risk harmful, and potentially fatal, electrical shock to personnel. In general, leaks should have been eliminated before the power tests are started but sometimes, and this is especially true in the case of pumps, a leak springs from the operating equipment during testing.

The best thing to do when a piece of electrical equipment gets wet it to turn it off immediately. Most pumps have an emergency stop switch and many types of equipment have an on/off button. Staff should be encouraged to take the safe course of action and shut the power off to the equipment whenever necessary.

It also helps to know the locations of the electrical circuit breakers in case the room where the emergency stop switches are flooded.

All electrical breakers should be marked before the electrical performance tests are started. All other known and recommended safety precautions and regulations should be followed.

Equipment Tests

For most of the equipment mentioned in Chapter 11 the facility manager should conduct a test when it is first installed to make sure it is operating correctly.

Equipment tests are performed according to the standards by which they are purchased. If a water softener is supposed to make 6,000 gallons of soft water between regenerations, a test needs to be conducted to confirm that 6,000 gallons are produced before the water starts to get hard again. If the water softener is supposed to regenerate automatically, the facility manager should check this out as well.

Equipment tests should be specified in the contract and are the responsibility of the construction contractor. The facility manager should be aware that the construction contractor will subcontract the purchase of a piece of equipment from a supplier. The supplier represents a manufacturer who designs and sells the equipment.

When an equipment test is conducted, the supplier usually has a representative of the manufacturer come out to the facility and conduct the test. The facility manager does not want to become entangled in any disputes between the various subcontractors, suppliers or middlemen but this may be unavoidable if there are problems with the equipment. Recalling the previous chapter's discussion on payment, the facility manager does not want to pay for equipment that does not work and consequently steps should be taken by the facility manager to assure that the tests are completed before the equipment is paid for.

Start-ups

When it comes to starting up new equipment, the facility manager benefits greatly from having his maintenance staff witness the performance tests. In this way, the maintenance personnel can meet with the manufacturer's representative and learn all they can in the short span of time available while the start-up and performance tests are done.

In addition, most equipment comes from the factory with an operations and maintenance (O&M) manual which is turned over to the facility upon completion. This manual is the manufacturer's instruction manual on how to operate its equipment. Care should be taken with the O&M manuals because they are one of the most important pieces of information for operations personnel after the contractors and manufacturers' representatives go home. The O&M manual provides valuable information in setting up a preventive maintenance program for the equipment (see Chapter 17).

It is also beneficial for the maintenance staff to be able to contact the manufacturer directly since spare parts or accessories will have to be purchased later on.

In most cases, equipment also has to be subjected to leak testing. Often, the manufacturer performs a leak test at the factory before shipping it, but sometimes during installation vessels and pipes get out of line and small leaks appear. These, of course, should be fixed before field testing.

Almost every piece of equipment has some type of control mechanism that makes the equipment self-regulating. The controls mean that staff only has to check on the equipment once in a while to make sure it is working correctly. The controls have to be set up initially and checked, and then the system can operate with minor troubleshooting.

One of the problems with start-ups is that it takes a few days for an entire system to "work out the bugs." During this time, pressures are fluctuating, power is being turned on and off and new people are looking at and "fiddling with" the new equipment. This activity leads to some unexpected headaches for the facility manager and it is a specialty field for many engineers and electronics technicians. Some like start-ups, and some do not.

Filter Equipment Tests

Equipment tests for filters are similar to the tests for water softeners: leak tests, controls tests and finally a full-performance demonstration. Water purification equipment must be tested to ensure it removes particles and fine materials, but no facility wants to put debris into the system just to be assured the filters will work. The test procedure should be written out in advance and the facility manager would then review the plan for the test, agree to it, and then the actual tests proceed.

Chlorinator Testing

Chlorinators and other water purification equipment requires lab test equipment to make sure the water has the right amount of chlorine in it.

Samples may have to be sent to the lab or a small test lab set up with the chlorinator just for the test. Ozone and ultraviolet tests are conducted with similar equipment.

Water Heating Equipment Tests

Water heating equipment tests are conducted in a similar way to the equipment performance tests, although additional measurements are taken of the fuel consumed during the performance test. The consumed fuel readings are used to calculate efficiency of the equipment.

The facility wants to keep a record of this test because as his water heating equipment ages, various energy loss elements affect the efficiency of the unit. With documentation of the efficiency from when the equipment was new, the facility can determine when to schedule tank cleanings and burner checks to keep the equipment operating at peak efficiency.

Disinfecting Water Lines

After the lines have been flushed and pressure tested, the lines may need to be disinfected to make sure they don't harbor pathogens (Chapter 4.) Usually new underground water mains and major repairs require treatment. Local above ground lines and short runs may have enough residual chlorine in the incoming water that any microbes would be killed. In some facilities, if the line is less than 5 feet to the tap and it is smaller than 3/4 of an inch, microbes will be flushed out during the flushing and flow test.

For large systems that must be disinfected, the most common chemical is chlorine. Other methods include ultraviolet light or ozone, but these methods are more complex. American Water Works Association Standard (AWWA) C651 is often used in contracts for constructing new water lines as a guide to disinfecting water mains with chlorine (Chapter 19.)

Under the AWWA Standard there are three methods for disinfecting water lines. They are: the tablet method, the continuous feed method, and the slug method. The tablet method is a favored method of treatment since tablets of calcium hypochlorite are added to the water to increase the chlorine to 25 milligrams per liter for 24 hours. In the continuous feed method, chlorine is added to the flow to bring the free chlorine level to 10 milligrams per liter after 24 hours, and in the slug method chlorine is added to give a concentration of 50 milligrams per liter for a minimum of 3 hours. If potable water was used, this increase in chlorine should adequately disinfect the system.

After disinfecting lines are flushed to rinse out the concentrated chlorine, checked to assure there is adequate chorine remaining (between 1 and 5 milligrams/ liter) and tested and verified free of contamination, the water lines are ready for service.

Test Results

The results of testing are a facility manager's tool to keep the equipment running in top shape. They protect the facility from liability and keep the maintenance force on its toes. Once the systems are in, tested and running, the facility manager has to make sure the systems are well-maintained. In the next chapter, we will review some tips for maintenance operations to keep the systems running smoothly and save the facility money in the long run.

Chapter 17
Maintenance

*T*hroughout this book, attempts have been made to indicate where maintenance can be easily facilitated during the design, operation or construction of a water system. The overall goal and objective of the system is providing good service to the customers. After design, maintenance is the key to the facility manager's success for a smooth operating water system.

ELEMENTS OF MAINTENANCE

There are essentially two kinds of maintenance activities. Preventive maintenance is checking the system, making sure it is running smoothly, lubrication, changing filters and other routine tasks. In addition to preventive maintenance, there are repairs that must be completed.

Repair projects, however, can be minimized if the preventive maintenance is successful. For example, checking a pump is preventive maintenance, while replacing it because it failed could be considered non-routine maintenance. Managing maintenance activities requires the skillful combination of routine preventive maintenance and properly scheduling and conducting non-routine maintenance.

The manager uses labor, materials and tools to perform maintenance, His success or failure depends to some extent upon the effectiveness of his use of these resources.

Labor

To successfully maintain a facility water system, the manager must bring together the right combination of labor. Labor, the people who work on the system, must be empowered with a strong sense of authority and responsibility. In a sense, they own it. The facility manager must turn over to the staff some of the authority to see that the work gets done. The manager must see to it that the workers are properly trained and that they have the necessary skills. The workers must know where all the valves and pumps are, and how to get to the equipment needed. In addition, they must have the confidence that they are capable of doing a good job. Pride and accomplishment plays a major role in the successful maintenance of any system.

The labor force in turn must have confidence in the management. That is, the workmen and women have to believe that if they tell the manager something is needed, the manager will see to it that it will be done. Often, a manager relies on foremen to manage the craft. Foremen are the sergeants of operation, responsible for daily prioritizing tasks, coordinating materials and labor activities.

Materials

Materials management is difficult given the modern state of technology and its continuous change. Manufacturers continuously and rapidly change their equipment and they are changing the salespeople even more rapidly. The dynamics of the industry make it difficult to keep the necessary repair parts in stock for a facility.

In addition, given the modern structure of the organization, the authority to purchase materials is often held at a high level—sometimes, even beyond the facility manager's control. For successful water system repair, the maintenance staff must have the correct part. They must know where it is needed, how to get it, and how to install it properly so that it works correctly the first time.

The question of how much spares to keep on hand is a difficult one. Some vendors sell cheap equipment and do not stock the necessary spare parts for repairing it. Others attempt to sell a huge spare parts repair kit along with the equipment when it is not really necessary.

This problem is compounded in a facility where there is little space for storing the spare supplies. In an ideal system, the worker knows exactly where the repair parts are stored, he is able to access them quickly, has seen them before and knows how to install them. If there is no warehousing of spares or if the warehousing is not very well organized, the worker cannot locate the repair parts and time is lost looking for them.

Tools

Many of the same things that can be said for spare parts can be said for tools. The maintenance staff must have an adequate supply of the right tools—wrenches, saws, drills, etc.—to fix and repair water system equipment. For a utility, tools would include large items such as digging machines. Many tools have a consumable element—for example drill bits break and backhoes must be refueled. These tool consumables must be accounted for and replaced in any water management system.

Also included in the category of tools are personal protective equipment necessary for performing the work safely. Personal protective equipment could be a welding hood for a welder or a respirator and gloves for patching paint coatings or doing tile work.

The successful manager brings these elements together in a cohesive pattern. He is successful in creating a system where labor, materials and tools are focused on the mission to maintain and repair the water system so that it consistently serves occupants well.

The major element of success for a facility manager is to optimize what is called "hands-on-tools-time" because this is the clearest evidence of where the three elements of labor, materials and tools come together. A facility manager should encourage hands-on-tools time, reward it when he sees it and chastise workers when he does not see it.

There are hundreds of excuses for not seeing hands-on-tools-time. Looking for parts, waiting for permission to shutdown, not having the right personal protective equipment, taking a break, talking with the boss, talking with the secretary, ordering parts and checking out tools are all necessary in the course of a work day. But loss of efficiency soon gets out of control even in well-run organizations. If the craftsman is waiting on parts, he can grab a broom and sweep, he can sharpen knives, he can clean paint rollers. There is always plenty of hands-on-tools-time that can be done.

WORK ORDER SYSTEM

The successful combination of the three elements of labor, materials and tools can be managed through the use of a work order system. Each repair, job or preventive maintenance inspection can be accounted for and tracked.

Many work order systems are commercially available that run on personal computer systems. Some are even available in the public domain, although they may be more difficult to find and operate than to purchase a marketed one.

Work order systems generate work order forms, keep track of equipment in a database, are capable of tracking and identifying trends, and record the number of labor hours assigned against the number of labor hours available.

The work order system starts with Preventive Maintenance Inspections and Preventive Maintenance Examinations, sometimes shown in abbreviated forms as PMI/PMEs.

When a new facility is constructed, all of the installed equipment—i.e., the pumps, the filters, the heaters, chlorinators and other items—comes from the manufacturer with an operations and maintenance (O&M) manual.

Inside the manual, there are recommended intervals for routine inspections. For example, a service manual on a pump may say to lubricate the bearing every 14 days.

In a work order system, the task to lubricate this pump would be generated as a work order. Work orders can be self-generated by the repairman or, in the case of large facilities, a special position called a work scheduler prepares the work orders and makes sure they are closed out when completed. A typical work order form is shown in Figure 17-1).

The other type of work orders come from the building's occupants. A call comes in for a repair. A work order is generated that, in effect, directs the repair work. These repair work orders are sometimes open-ended, since the total amount of work, materials, and tools is unknown until after the repair work has been assessed.

The work order should estimate the time required to complete to work and it should indicate the tools necessary to perform the work. In addition, many facilities add a small tools charge to

Work Order

Date _____   Equipment Tag No. _____

Number _____   Man-hours Assigned _____

Scheduler _____   Personnel Assigned _____

Tools Required   _____   _____

                      _____   _____

                      _____   _____

Man-hours Used _____

Consumables Used   Fuel_____   Gallons

                      Vehicles _____   Miles

                      Tools/Bits/Blades _____

                      Other _____

Parts Required   _____

                      _____

                      _____

Location of Parts   _____   Inventory

                      _____   On Order   _____

Tests Required   _____   Due In   _____

Tests Completed _____ Date _____

Signature of Test Witness _____

Occupant Caller _____Phone   _____

Signature of Employee Completing Work Order _____

Comments   _____

                      _____

                      _____

Figure 17-1. A typical work order form used for maintenance and repairs.

help budget for small tool replacement and consumables. The facility manager decides how much of this responsibility to delegate to the foremen or to the individual laborers.

Inventory

In theory, a good work order tracking system can be tied to a good inventory system. This way, if the work order requires the use of pump seals, the inventory system automatically subtracts one set of seals from the inventory.

However, a system like this almost never works well in a practical application, they are expensive to establish and they break down when parts are back-ordered or are no longer available.

They do work well for specialized types of projects, but for a large water system that includes multiple subsystems like hot water, wastewater, water treatment and storage, they have not proven effective.

Most successful facilities have a separate inventory control system and delegate the repair parts decisions to the labor force. The craft decides which repair parts are necessary and requests them from inventory. A warehouse person issues the parts and records the changes in stock levels. When the level runs low, new spares are ordered.

The process of inventory control is a difficult one for any facility manager. The dollar value of the stock in spares can be significant and higher management cannot understand the need for hundreds of thousands of dollars sitting idly in inventory.

The value of tying inventory to the work order system is that it will show if the needed part was available, or if it was not, and allows for an evaluation to be made of the advantage of having the material in spares inventory versus hoping the spares are available at an equipment vendor's downtown.

In addition, the facility manager should recognize that some elements are critical to the operation of the facility.

For example, a large pump may be used in the system to pump water into or out of a reservoir. This type of pump may not be readily available, and the cost to the operation of being without water may be insignificant compared to the cost of a large pump in inventory.

It also helps if the facility manager has these types of numbers at his fingertips in order to justify inventory of special large end item repair parts.

An example of this might be a large pump, say in the range of 150 horsepower. A pump this size could cost as much as $10,000, but if the pump supplies water for cooling the factory, and the factory costs $50,000 a day to operate, then the day saved by having the pump in stock in inventory is well worth the investment, as opposed to having to wait three days to have the pump flown in.

STAFFING FOR MAINTENANCE

Once all of the required Preventive Maintenance (PMI/PME) is known, it is a simple matter to calculate the required staffing level. The routine service items from the vendor's equipment manuals daily, weekly, monthly, quarterly, semi-annual and annual inspections—will total a number of man-hours. When completed, the facility manager knows about how many hours are needed to keep the systems running smoothly and with little interruption. Given all of the equipment and all of the inspections, the facility manager can then make a trial estimate of the size of the staff needed.

The average worker is paid for 2,080 hours in a work year. With time off for vacation and illness, this number drops to 1,896 hours (two weeks of vacation = 80 hours, one week of illness = 40 hours, eight holidays = 64 hours).

The repair work called in from customers is a function of the age of the facility and the number of customers. For a new facility, the manager can start out with a staff to cover the PMI/PME. There is usually enough extra time to accomplish a few call-in orders.

As the facility ages, adjustments are made. In addition, the facility manager can outsource some of the work if it becomes too burdensome.

Once the facility manager has a work order system up and running, trends and analyses can be used to look at how successful he is with his maintenance force. Trends include the number of work orders in a month.

There will always be a backlog of work orders and the facility should have a mechanism for prioritizing them. However, work orders must be completed on schedule because the customers will lose confidence if they are not. Once confidence begins to be lost, the maintenance mission is in jeopardy.

The manager should also provide for some kind of check/audit of the work order system to verify there are not too many excess hours assigned. Audits of work order systems can be performed by skilled facility management firms on a contract basis.

MAINTENANCE TIPS AND SHORT CUTS

Here are a few tricks of the trade learned from years of water system management and service. At least some should prove useful to the facility manager or members of his staff in scheduling work and performing maintenance.

Hot Taps

One of the more difficult tasks in working with a water system is attaching a new line to an existing one. Normal practice would dictate the existing line be shut off and drained, the new line attached, and then the system refilled and sterilized. In many cases, it is not practical to shut off and drain an existing line.

In this case, a hot tap can be employed. A hot tap is a process whereby a new line is attached to an existing, operating line without draining the existing one.

Preparing a hot tap can be a delicate operation, because if it is not done properly, the existing operating line has to be shut down if the hot tap fails.

Since the decision to install a hot tap is somewhat predicated on the inconvenience of shutting the line down, the failure of the hot tap defeats the purpose of the hot tap in the first place.

To successfully hot tap an operating line, the location of the new tap and its size are determined. Pressures and flows for the new line are checked and verified to be compatible with the design of the tap. Next, the service line is prepared for the hot tap. Preparation depends upon the location and type of line. If it is buried, preparation is made by excavating and cleaning it. If it is a hot water line, the insulation is removed and the line cleaned.

The equipment for a hot tap is sometimes commercially available. For water lines, a hot tap is sometimes called a saddle tap. Figure 17-2 shows a saddle tap installing an instrument probe.

Basically, the hot tap wraps around the pipe with clamps or bands. The mechanism for tapping the pipe consists of a drill or punch that uses either threaded punch or has an opening for a drill bit. For the latter, the bit extends through a membrane around the drill bit to reduce the leakage once the tap is complete.

After the punch or hole is drilled in the pipe, the punch is extracted. Depending upon the design, a valve that has been incorporated with the hot tap is closed. The new pipe is constructed downstream from the valve.

The design of saddle taps and hot taps depends to some extent upon the design of the pipe to be punched. For larger pipes, say above about 6 inches in diameter, the design of the taps depends upon strengths and stresses generated by the tap and clamps.

Caution should be exercised when installing a hot tap because some main line failures have occurred when the tap was improperly installed and the main line completely failed, causing a major flood and an outage in the water line. In addition, there have been significant failures because a hot tap was installed as a short-term measure with the intent to go back and make it more permanent when the main line was

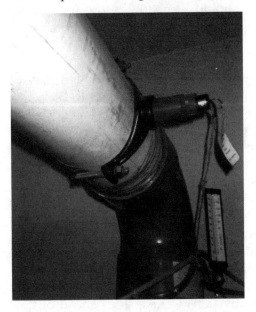

Figure 17-2. A saddle tap installing an instrument probe. Saddle taps can be used to attach a new line to an existing one without shutting off and draining the existing line.

shut down for maintenance. Years later, corrosion caused the clamps of the hot tap to fail, resulting in a failure of the main line.

For some small lines under low pressure—3/4-inch or smaller and less than 25-45 psi—it is possible to work the entire operation "wet" provided there is a place for the water to run that does not damage the facility. The pipe is cut, allowed to spray, and the new joints and valves added working around the pipe while spraying. This type of tap is not often done, and usually when it is done it is the result of a mistake by someone somewhere, but it still can be done. Advance planning is the best tool to prevent this type of work being done on an emergency basis.

Since wastewater lines are not usually pressure lines, a hot tap is not used. A simple tee is cut into the line instead and the wastewater allowed to drain while the new tee is installed. Hot taps for supply or wastewater have the potential to be "wet" operations and equipment such as mops, buckets, pump drains, raincoats and safety glasses should be ready before the hot tap takes place.

Utility Shutdown

Depending upon the service and the time of day, it may be more attractive to shut down the system rather than to try to hot tap it. For the facility manager, a utility shutdown should be coordinated in advance with the occupants and if they are told of the advantage of the shutdown ahead of time, most occupants will cooperate.

The facility manager should exercise caution when shutting down water systems because the occupants could be in a situation where loss of water could contribute to a serious problems for them. Fortunately, water and wastewater utility shutdowns are not as critical as power outages, but the same principals apply. Some facilities, notably hospitals, require signatures of the various services before the facility manager can turn off one or more utilities. This allows the various divisions to take steps ahead of time to mitigate the impacts of the water system shutdown.

Freeze Plug

Commonly used when a valve for shutoff is not available, a commercial electric blanket that works on refrigeration principles is wrapped around a pipe. When the blanket is turned on, it freezes the pipe along with the water inside, plugging it. Downstream the pipe can be cut or tapped while the plug is in place. When complete, the cold blanket is turned off and as the ice melts, the water can flow again. A few problems with freeze plugs are that the plug can damage some kinds of pipe—i.e., split it because the ice expands. Another problem sometimes encountered is that the freeze plug does not hold—that is, it slips under the pressure of the water upstream and blows out the end of the cut pipe.

Balloon Plug

Similar to the freeze plug, a balloon plug is sometimes used. A small hole is drilled into the pipe as a hot tap and a rubber or plastic balloon is inserted through the hole and into the pipe. The balloon is filled with air and chokes off the pipe flow.

Unplugging

Fortunately, water supply pipes rarely get plugged because of the filtering and quality requirements. As supply water piping is pressurized, plugging reduces flows at faucets or toilets and results in an occupant complaint. Locating the plug is relatively simple, but a system shutdown is necessary to remove the plug. Sometimes on smaller lines, the plugged section is identified and hot taps are located on either side of the plug, new pipe is run between the hot taps and the plugged pipe is abandoned in place.

Wastewater pipe is known to plug more often because of the solid materials carried in the flow. In addition, stringed items such as dental floss contribute to plugging. Fortunately, wastewater piping is not pressurized and clogs are removed with wire coil called a snake. Many plumbing firms are willing to be called out on a 24-hour basis and snake sewer lines for a fixed or hourly fee. Depending upon the number of incidents at a facility, this type of service can be contracted out or the staff can buy the necessary equipment and use it when needed.

Plugging wastewater lines can be traced to a repeated practice. At one facility, the sewer lines were old and ran close to a large tree. In the spring, the sewer line always backed up until the facility manager was able to locate a chemical that discouraged root growth. The chemical was put into the piping and the roots from the tree ceased to be a problem. Facility managers should be careful when adding a root herbicide to the wastewater, however, since it can also affect operations at the sewerage treatment plant.

CAMERA INSPECTION

For buried outside waste water lines, it is common practice to conduct a camera survey of the lines every five years or so. This type of service is usually contracted to a few firms that specialize in this type of service.

A Word of Caution About Contracting Out

Many unions and labor organizations are opposed to having services that have historically been done in-house contracted to people or companies from outside the facility. Most of these facilities operate under a contract that exists between labor and management and agreements have been made ahead of time as to what services are going to be contracted and which are not. If the craft have been given a strong sense of ownership, they will be just as interested in contracting out special services as the management. However, if it appears that management intends to contract out services such that people's functions and responsibilities are threatened, the facility manager will have tough sledding. Usually, the facility manager does not want to get involved in a labor battle. The facility manager should try to determine if this concern exists before making a decision to contract out services.

Figure 17-3 shows a typical camera and Figure 17-4 shows a typical truck rigged for the operation.

A small camera is either pulled through the wastewater pipe or the camera itself is mounted on a small tractor that crawls through the pipe. The camera provides a cable feed to the suppliers truck where the camera view is videotaped.

Figure 17-3. A typical camera used for pipe inspections. Courtesy: Aries Industries, Inc., Sussex, WI.

Figure 17-4. Typical truck rigged for camera inspection. Courtesy: Aries Industries, Inc., Sussex, WI.

The survey can analyze to tapes with data and a report, or the tapes can be provided to the facility manager alone. Often, the service company is able to provide, as digital input directly on the videotape, the information indicating where the camera is positioned. The service is useful in locating leaks, breaks and other operational problems with sewer lines.

LEAK DETECTION

Similar to the camera inspection service, special companies will provide for leak detection as well. Leaks present in a system account for reduced pressures and flows and lead to increased costs. Sometimes, these leaks can go undetected because the piping is below ground or located in an area where it cannot be physically inspected. Leaks can also be detected using test methods discussed in Chapter 16.

The special leak detection firms use sound detection devices to listen for the leaks. Other methods include addition of inert materials to the water that can be detected but are not considered contaminants. If the system can be shut down for an extended period and drained, a tracer gas can be used to pinpoint the source of leaks. Tracer gas can be sophisticated gas such as helium or sulfur hexafluoride or even a simple odorant.

The odorant in fuel gases is called methyl mercaptan, but its use is somewhat regulated since the odor is the odor of leaking fuel gas, and confusion may arise because occupants will suspect a fuel gas leak.

Pipe Spools

In some facilities, valves are ordered and are delivered late, or valves have to be removed for service because they leak. Since these valves are sometimes expensive, the facility does not want to go to the expense of purchasing a second valve while the first is repaired.

In this case, a small pipe spool piece is made up, exactly the length and diameter of the valve. The line is plugged on either side using freeze or balloon plugs, and the valve removed. The spool is put into the place of the valve while the valve is taken into the shop and overhauled.

While the valve is being repaired, the plugs are removed and the line placed back into service. When the valve repair is complete, the same process is done in reverse to remove the spool and put the valve back in.

This only works if the valve to be changed is a service valve used to isolate a portion of a system. If the valves constantly regulate flows, then another arrangement has to be made by renting, borrowing or purchasing a spare valve. The size and bolt patterns must match, of course.

Chapter 18

Managing Water Personnel

*F*rom routine maintenance and operations to renovating or building a water system, the facility manager will coordinate both his own staff and a diverse range of trades and professions. In this chapter, we will review the key people the facility manager will deal with on water projects, as well as discuss ways the facility manager can keep abreast of new products, legislation and technology.

THE FACILITY MANAGER

The facility manager manages both the people who work on the water system and the system itself. His role is to integrate, the water system, his staff and the necessary repair equipment to provide water to occupants that is safe and cost-effective. If the water is used for drinking and bathing, it should meet cleanliness standards. For other applications, it should meet the intended purpose. Wastewater should drain freely to the sewage treatment plant or to a sanitary sewer where it can flow to the sewage plant.

Stormwater runoff should carry over to storm drains or stormwater ponds. The ponds should hold the water, allow it to be released slowly so it can be used by others, or allow it to be reused by the facility or wasted into a river in a safe environmentally sound manner.

The facility manager will pay for water used, pay for electricity to pump the water, pay plumbers and pipe fitters for working on the system. In addition, the facility manager will pay for laboratories to test the water to make sure it is safe (or he may have a laboratory on site where he does his own tests) and he will pay for sewage treatment.

To accomplish each of these goals, the facility manager will work with people. This chapter discusses some of the people the facility manager will work with, and their approximate levels of training and expertise.

Note that a discussion of this type has a margin of error because we are talking about people and not about things. Generally speaking, water professionals like what they do. They have achieved success and are proud that their work is used and depended upon by many people. A good facility manager recognizes this trait and seeks to create an environment where the skills of the water system professionals can shine.

WATER PROFESSIONALS

Key water professionals the facility manager will come into contact with include designers, planners, lab technicians, plumbers, pipefitters, contractors, plant operators and representatives of equipment manufacturers.

Designers

Water system designers include engineers, technicians and estimators. Employees or consultants with this expertise are usually well educated, the engineers with a four- or five-year degree, the technicians with a four-year science degree or a two-year associate degree from technical school.

These professionals will write contracts and prepare drawings from master specifications and drawing guidelines. The facility manager relies upon their creativity, technical expertise and judgment for a reliable product.

In addition to the years of training, some of the engineers will be registered. Registration requires application through a state board, a degree and passing both a fundamental and practical examination. Registered engineers use an embossed stamp, like a notary, to stamp drawings to certify the project meets professional codes or standards.

Many cities and governments require that public water systems be designed by qualified designers. Usually, this requirement means registered professional engineers. The licensing state will provide a list of engineers to facility managers who request them. Registered professional engineers are subject to rules relating to conflicts of interest and competition.

Designers' tools include computers and software, engineering texts and design manuals, materials for preparing drawings which include Computer Aided Design (CAD or ACAD) software, computers, monitors and plotters. Other tools include typewriters and text generating materials such as printers, copy machines and a drafting board where drawings can be laid out and edited.

Planners

Planners can be technicians and licensed or unlicensed engineers, or they can be members of the facility manager's staff. Planners do not have to be licensed, although they can be. Some engineering firms specialize in planning and offer it as a service as well as detailed design work.

Planning is best done by employees of the facility since they are most familiar with the goals and objectives of the facility. In general, contracted planners tend to focus on one area they have had success with in the past. Some specialize in growth, others in conservation. Planners' tools include telecommunications equipment, text-generating equipment (computer word processors, printers and copy machines), computers that prepare mathematical models, and texts and manuals. Planners usually subscribe to a magazine about planning and project management.

Lab Technicians

Lab technicians collect and analyze facility water samples. Samples do not necessarily have to be collected by lab technicians themselves—the facility manager can arrange to have one of his staff trained to take the samples. Once the samples are collected, they are analyzed in a laboratory.

Depending upon the complexity of the tests, lab analysis can be very expensive. See Chapter 4 for the time and costs typically involved in performing various tests. Lab technicians carefully measure and analyze the results.

Lab personnel can have up to eight years of college and several levels of science degrees. Lab technicians have 2-4 years of college. Some work shifts and odd hours depending upon the needs of the laboratory.

Lab tools include glassware for mixing samples; ovens and refrigerators for storing and drying samples; storage cases for lab chemicals to prepare samples for examination; and computers for generating reports and data. Some sophisticated laboratory instrumentation includes gas chromatographs, scales and chemical dyes. Microscopes are still used along with a fume hood for mixing chemicals. Drains in a laboratory must be suitable for handling the chemicals poured down them.

Not all laboratories have the necessary tools and equipment for analyzing every alternative. Labs share special equipment. Most water labs have to send samples to another laboratory for chemical analysis of one type or another. The American Water Works Association publishes the standard test methods for analyzing water samples.

Plumbers

Plumbers enter a strict apprenticeship program where they receive on-the-job training for up to four years. In this training, they learn codes and standards and various pipe cutting, fitting and fabricating techniques. Plumbers learn construction jobsite housekeeping and safety as well. They are trained to safely and efficiently operate the tools used in pipe fabrication and installation.

Tools used include saws, welders, jackstands, tape measures, level, square, plumb bob, drills, hammers, wrenches (pipe, crescent, socket, box and open end) gloves, goggles, eyeglasses, heavy boots, hard hats, helmets and face shields. For scaffold work and work above ground, safety equipment includes belts, harnesses, ropes, pulleys, chains, chain falls, wrecking bars, pry bars. On many jobs, plumbers will need to be assisted by heavy equipment such as backhoes, front end loaders and cranes. Plumbers also install the china fixtures in rest rooms as part of finish work.

Pipefitters

Pipefitters' duties are similar in many ways to plumbers, and in some areas these trades are interchangeable. Pipefitters usually install larger pipes in industrial facilities and include welders.

However, not all pipefitters are qualified welders.

Pipefitters are trained in an extensive apprenticeship program that includes training in codes, standards, fabrication and jobsite housekeeping and safety.

Tools are similar to those of plumbers, except they are larger because of the larger sizes of the pipes.

Contractors

Water system contractors include a broad range of people—however, the three most important are mentioned here.

The owner of the contracting company owns the shops, equipment, materials and tools. He decides which jobs the company will bid and usually has the final say in what the bid is. He is well rewarded for his effort and is usually an entrepreneur. Training can have been obtained through university or trade study or it can be the result of years of experience. As an owner, he is the manager of one or more of the company job elements.

The superintendent is usually hired by the owner to run a job at a facility. Sometimes, the owner is the superintendent and rare is the company where the owner has never been a superintendent on one or more jobs. The superintendent manages the individual foremen on the job and manages resources at the jobsite. The superintendent will handle construction planning of his company's portion of the work and will coordinate material deliveries, storage, billing, materials and tool control. The superintendent is responsible for jobsite safety, cost and damage control.

The foreman manages one element of the job or one crew. The foreman is usually one of the more experienced of the craft who functions as a leader of the crew. Depending upon the size of the task, the foreman can supervise from one to 10 crewmen and a few elements of equipment. The facility inspector will inspect work at the foreman level.

Plant Operators

Water supply and wastewater treatment plant operators have similar training backgrounds and education. Most have two years of college education beyond high school and senior operators will have a four-year degree. For utility plants, the plant manager will often have an engineering degree but it is not always a requirement. Many plant operators will have a trade background as a plumber or machinist.

Tools used are similar to plumbers' tools. In addition, a large utility plant sometimes uses the skills of an employee with an electronics background to service and troubleshoot the instruments used to control the plant's equipment. As electrical operating components become more sophisticated, there may be a need for a computer programmer familiar with instrumentation and automatic controls.

Equipment Vendors

With the rapid changing of the state of water system equipment, it is a good idea to allow the staff some access to vendors. In this way, they have the opportunity to see new products and tools that have the potential to make the work go more smoothly.

Equipment vendors, however, perceive sometimes that a commitment is being made to purchase the equipment and this can pose a difficult problem. For example, if the vendor offers use of a tool and it gets broken during the course of use, who will pay for it? In addition, craft sometimes try out a new tool and become convinced of its usefulness but, when the order comes in to purchase it, the facility manager decides not to buy it because he did not know the craft was using it. This leads to a loss of morale because the staff used the equipment to save money but were not told of the cost of the item. The facility manager, on the other hand, knows how much the item costs but really does not know what it can or will do.

In recent years, equipment vendors have become more service-oriented. They have learned that just making a sale is not as much of a success in business as repeat business and therefore try to concentrate on repeat business as well as the initial sale. By and large, equipment vendors, while familiar with the product, are not as technically trained as engineers. Some have a four-year college degree but others have less schooling.

The facility manager, recognizing what has been said here, should establish a clear policy relative to contact and use of vendor-supplied products. Whatever the policy is, he should make sure all of the facility employees stick to it and make sure the employees notify the vendors of the policy as well.

SAFETY

One of the roles of the facility manager is to assure worker safety. The law requires that certain requirements for worker safety be met. Worker safety and the Workers Right To Know Law were touched on briefly in connection with Chapter 11 where there was a discussion of Material Safety Data Sheets (MSDS).

Discussed here will be three of the main elements in OSHA 1910 Worker Safety Laws. These three elements are the Workers Right To Know Law (MSDS), personal protective equipment and confined space entry requirements. The three discussed below are typical elements encountered by employees working on water systems.

Material Safety Data Sheets

Incorporated in the body of OSHA 1910 is the Workers Right To Know Law. This requirement basically states that the worker has the right to know what chemicals are in the workplace, what the hazards of working with those chemicals are, and what he can do to protect himself from their hazards.

Every chemical manufacturer is required to supply an MSDS with the chemicals. The MSDS indicates what the chemicals are, what the hazards are, and what the workers can do to protect themselves from the chemical. A sample MSDS for chlorine, a common chemical used in water treatment, is included in Chapter 11.

Personal Protective Equipment

The employer, in this case a facility manager, has the obligation to supply the necessary personal protective equipment whether or not it is identified in the MSDS. For employees working with water supply or wastewater systems, personal protective equipment includes face shields, gloves, respirators, boots, hard hats, ear plugs and all types of wet weather gear. These items are intended to prevent eye, hand and respiratory injury.

In general terms, management or the employee should identify the hazard and provide the necessary equipment to mitigate the hazards. Hazards can be physical—such as a pinch point working with heavy equipment; chemical—such as working around chlorine or other disinfectant; electrical, such as working around pump motors; or biological—such as microorganisms in raw water supplies.

Finally, if there is a hazard identified with a confined space such as a tank, where the employee gets inside the vessel, to perform some type of maintenance, the facility should have a written procedure for confined space entry (see the questionnaire in the side-bar). Confined space entry procedures require in essence that the worker have a plan for rescuing someone before they enter the confined space. Usually this means there is a spotter who watches them from outside the confined space and that the space is adequately ventilated to make sure there is no asphyxiation risk.

MANAGING SHIFT WORK

For the facility manager in charge of round-the-clock operations, shift work poses a unique set of problems. Perhaps one of the most difficult problems is knowing when and what decisions to delegate and recognizing that not each person on shift has equal capability. Most organizations pay extra for nights, Sundays and holidays, and the facility manager should see to it that the payment for shift differentials is fair and that the rotations among the staff are fair. If one employee wants to work nights only, for example, this could create jealousy from other staff members because of the extra pay involved. Another concern is when placing women and men on the same late shift and being sure there is no sexual harassment in the middle of the night.

Shift work should be scheduled far enough in advance to allow the workers to plan their activities around the work schedule. In general, this scheduling is at least five to seven days and more commonly it is two weeks to 30 days.

In spite of the manager's effort to schedule, there will be times when one of the staff is not able to come in due to illness, death in the family, etc. Since the odd shifts are usually thinly covered anyway, it will be necessary for one or more to work late or to come in early to pick up the balance. The shift manager should recognize this in the staffing budget. There are 8,760 manhours in a year (24 hours per day x 7 days per week). If the facility

Confined Space Entry Procedures

1. Have confined spaces on the facility been defined?

2. Hove confined spaces on the facility been noted on a map or other drawing?

3. Have workers been trained to recognize confined spaces?

4. Have workers who enter confined spaces been trained in rescue procedures?

5. Is the necessary equipment available for workers who make confined space entries?

6. Have the confined space entrants been trained on the confined space entry rescue equipment?

7. Is a buddy system in place where the workers in confined spaces are watched by trained rescue personnel?

8. Is there a communication system in place where the rescue personnel can communicate with both the entrants into the confined space and the rest of plant personnel in the event a problem develops and a rescue is necessary?

Source: Occupational Safety and Health Regulations for Worker Safety.

manager has been successful in inspiring a strong sense of loyalty to the system in the work force, the staff will help keep the system covered as needed.

KEEPING ABREAST OF TECHNOLOGY

The facility manager must try to keep abreast of the rapidly changing technology and here are a few techniques to help the facility manager get ahead and remain ahead.

Magazines

For a water manager, several magazines are available to keep informed on issues and technology. The American Water Works Association (AWWA) has several publications that serve to keep the facility manager abreast of issues relative to water supplies. With a membership in the organization, the facility manager receives *Journal* a professionally written and prepared magazine that provides legislative information and four or five technical articles in each issue. The technical articles may be too complex for the facility manager's needs since these are usually scientific papers discussing state of the art technology. In addition, AWWA publishes *OpFlow*, a newsletter for water supply system operators which discusses some of the recent legislation along with operational tips for water system managers.

Another commercial publication is the *Water and Wastes Digest* which is dedicated primarily to vendors of new items of equipment. The magazine usually publishes semi-technical articles of interest about new technology.

The American Society of Civil Engineers publishes *Civil Engineering*, which is primarily geared toward design with some emphasis on large construction projects.

Each of these organizations is referenced in Chapter 19.

Associations

Listed in Chapter 19 are a number of trade groups and associations that represent numerous interests within the water management spectrum. Contact with one or more of these associations will provide the facility manager with information about new products, rules, codes and standards. In addition, several associations provide information on costs of equipment and staffs. Associations also hold quarterly and annual meetings to discuss association business. Most associations try to coordinate these annual meetings with a small amount of recreation, such as golf and sight-seeing to break up the intensity of the work conducted at association meetings.

Teleconferencing

Somewhat highly specialized, teleconferencing is a way to be informed of association business and water management techniques. Teleconferencing is often sponsored by one or more of the associations and consists of a panel discussion with call in questions. The problem with a teleconference is that they usually require the meeting to be held at a central site, sometimes a large hotel or host facility. It may be inconvenient for the facility manager to leave his place of work to go to the teleconference. Usually, the presenters of the teleconference tape the discussion. This way, the facility manager can rent or purchase the videotape and watch it at his leisure. However, it takes some discipline to watch a teleconference videotape because the speed at which the information is distributed is rather slow. A book such as this one is usually a much better source of information.

Correspondence

The facility manager, if he makes himself known that he is available to others, will discover a huge amount of correspondence, some of which may actually prove useful. There is a number of seminars, classes, home study courses and consultants that offer services to the facility manager because he is the decision-maker who has the authority to disburse funds.

Other Sources

Finally, there are a few other ways in which the facility manager can keep abreast of technology. This would include personal contacts which is probably one of the most valuable. The facility manager may run across a problem which he does not see every day, but which one of his business associates sees often. A phone call, well timed, can provide an excellent source of information, and for the facility manager, it will be timely and accurate. It takes time and energy for the facility manager to remain current in today's rapidly changing world but the rewards are plentiful. Quite often the right information at the right time can save the facility a considerable amount of money.

Chapter 19
Trade Groups and Associations

*A*s water management is a complex field of engineering, the facility manager needs to rely upon the advice of specialists, The trade associations listed in this chapter provide information in a wide variety of areas, including magazines and newsletters, training and sample specifications.

TRADE ASSOCIATIONS

The below list is not all-inclusive; if any association has not been mentioned here, please write to the publisher and the appropriate information will be included in future editions of the book.

American Fire Sprinkler Association
9696 Skillman Street, Suite 300
Dallas, TX 75243-8264
Phone: 214-349-5965
Fax: 214-343-8898
e-mail: afsainfo@firesprinkler.org
website: www.sprinklernet.org

A non-profit international association representing open-shop fire sprinkler contractors dedicated to the educational advancement of members and the promotion of the use of automatic fire sprinklers.

301

American Institute of Architects
1735 New York Avenue, NW
Washington, DC 20006
Phone: 800-AIA-3837
Fax: 202-626-7547
e-mail: infocentral@aia.org
Internet: www.aia.org

American National Standards Institute
25 West 43rd St., 4th Floor
New York, NY 10036
Phone: 212-642-4900
Fax: 212-302-1286
Internet: http://www.ansi.org
e-mail: into@ansi.org

American Society of Civil Engineers
1801 Alexander Bell Drive
Reston, VA 20191-4400
Phone: 800-548-2723
Fax: 703-295-6222
Internet: http://www.asce.org

One of the oldest of the professional engineering societies, ASCE publishes journals and technical publications subject to peer review about water supply, water resources, water planning and water and wastewater quality management.

American Society of Mechanical Engineers
Three Park Avenue
New York, NY 10016-5990
Phone: 973-882-1167
Toll Free: 800-843-2763
e-mail: infocentral@asme.org
Internet: http://www.asme.org

Similar in organization to the American Society of Civil Engineers, the ASME promotes the efforts of mechanical engineers in the design of pumps, machinery, and in boiler and pressure vessels. ASME publishes standards for piping and for heating equipment.

American Society of Heating, Refrigerating and Air-Conditioning Engineers, Inc.
1791 Tullie Circle, NE
Atlanta, GA 30329
Phone: 404-636-8400
Fax: 404-321-5478
Internet: www.ashrae.org

ASHRAE is the professional society for air conditioning. Its standards are nationally recognized and ASHRAE also publishes standards for hot water sizing and heating.

American Society for Testing and Materials
P.O. Box C700
100 Barr Harbor Drive
West Conshohocken, PA 19428-2959
Phone: 610-832-9585
Fax: 610-832-9555
e-mail: service@astm.org
Internet: www.astm.org

American Water Works Association
6666 W. Quincy Ave.
Denver, CO 80235
Phone: 303-794-7711
Fax: 303-794-3951
Publications: 800-926-7737
e-mail: mparmlec@awwa.org
Internet: www.awwa.org

AWWA is a non-profit organization that promotes standards for water quality, plumbing and piping. AWWA is the principal association of utility water managers for large and small communities.

Building Officials and Code Administrators International
4051 W. Flossmoor Rd.
Country Club Hills, IL 60478
Phone: 800-214-4321
Fax: 708-799-4981
e-mail: codes@bocai.org
Internet: www.bocai.org

Construction Specifications institute
99 Canal Center Plaza, Suite 300
Alexandria, VA 22314-1791
Toll Free: 800-689-2900
Fax: 703-684-8436
e-mail: membcustsrv@csinet.org
Internet: www.csinet.org

Studies the interaction of various construction disciplines and the
standardization of construction documents, conducts educational
programs and sponsors local, regional and national conferences
and trade shows. Prepares master construction specifications for
all types of building construction work.

Copper Development Association, Inc.
260 Madison Avenue
New York, NY 10016
Phone: 212-251-7200
Fax: 212-251-7324
Internet: http://piping.copper.org

The Copper Development Association, Inc., promotes and develops copper and brass pipe and fittings for the plumbing industry.

Ductile Iron Pipe Research Association
245 Riverchase Parkway East, Suite D
Birmingham, AL 35233-1856
Phone: 205-402-8700
Fax: 205-402-8730
Internet: www.dipra.org

**Foundation for Cross-ConnectionControl and
Hydraulic Research**
University of Southern California
Kaprielian Hall 200
Los Angeles, CA 90089-2531
Phone: 866-545-6340
Fax: 213-740-8399
e-mail: fcchr@usc.edu
Internet: www.usc.edu/dept/fcchr

Instrument Society of America
67 Alexander Drive
P.O. Box 12277
Research Triangle Park, NC 27709
Phone: 919-549-8411
Fax: 919-549-8288
e-mail: info@isa.org
Internet: www.isa.org

**International Association of Plumbing
and Mechanical Officials**
5001 E. Philadelphia St.
Ontario, CA 91761
Phone: 909-472-4100
Publications: 800-85-IAPMO
Fax: 909-472-4150
e-mail: iapmo@iapmo.org
Internet: www.iapmo.org

Publishes the Uniform Plumbing Code and the Uniform Swimming Pool, Spa and Hot Tub Code as well as other materials relating to model codes and training. Maintains a complete plumbing product listing service, which includes product testing, compliance, inspections and a test laboratory. Develops standards, presents educational seminars on plumbing and tests journeymen, and certifies inspectors.

International Conference of Building Officials
5360 Workman Mill Road
Whittier, California 90601-2298
Phone: 800-423-6587 (x3252)
Internet: www.icbo.org

Develops and publishes model building codes and standards, including the Uniform Mechanical Code and the Uniform Building Code. Publishes books and technical manuals about building technology and conducts training and seminars.

National Association of Corrosion Engineers
1440 South Creek Drive
Houston, TX 77084-4906
Phone: 281-228-6200
Fax: 281-228-6300
Internet: http://nace.org

NACE educates members of the public and technical professions about corrosion and materials performance and protection. NACE also works to find better ways of addressing safety, life of materials, and designs for corrosion prevention and control.

National Association of Plumbing-Heating-Cooling Contractors
180 S. Washington Street
P.O. Box 6808
Falls Church, VA 22046-2919
Phone: 703-237-8100
Toll Free: 800-533-7694

This is an association of plumbing, heating and cooling contractors dedicated to the promotion, advancement, education and training of the industry for the protection of the health, safety and comfort of society and the environment.

National Fire Protection Association
1 Batterymarch Park, P.O. Box 9101
Quincy, MA 02269-9101
Phone: 617-770-3000
Toll Free: 800-344-3555
Internet: www.nfpa.org

The NFPA produces fire safety codes, standards, handbooks, training and safety materials. Conducts seminars and training for fire protection designers, firefighters and other safety professionals.

National Fire Sprinkler Association
P.O. Box 1000
Patterson, NY 12563
Phone: 845-878-4200
Fax: 845-878-4215
e-mail: info@nfsa.org
Internet: www.nfsa.org

The NFSA promotes the manufacture and installation of fire sprinkler systems and fire sprinkler devices. The NFSA also promotes the recognition of the fire sprinkler industry as a unique identity. Conducts various educational programs and seminars. Publishes *Sprinkler Quarterly* and assorted guides and information pamphlets.

National Sanitation Foundation
NSF International
P.O. Box 130140
789 N. Dixboro Rd.
Ann Arbor, MI 48113-0140
Phone: 734-769-8010
Toll Free: 800-NSF-MARK
e-mail: info@nsf.org
Internet: http://www.nsf.org

Non-profit organization providing programs on public health and environmental quality. Develops and maintains consensus standards, tests and certifies products, inspects production facilities, registers quality systems and conducts special studies.

National Spa and Pool Instutite
2111 Eisenhower Ave.
Alexandria, VA 22314
Phone: 703-838-0083
Fax: 703-549-0493
e-mail: cdigiovanni@nspi.org
Internet: http://www.nspi.org

This institute publishes standards and promotes swimming pool and spa safety. It is comprised of public health officials and of pool manufacturers and swimming pool contractors.

National Swimming Pool Foundation
10803 Gulfdale, Suite 300
San Antonio, TX 78216
Phone: 210-525-1227
e-mail: evnspf@aol.com
Internet: http://nspf.com

A non-profit educational organization that initiates and supports research to improve aquatic facility safety, design, construction, operations and management.

Plastic Pipe and Fittings Association
800 Roosevelt Road
Building C, Suite 20
Glen Ellyn, IL 60137
Phone: 630-858-6540
Fax: 630-790-3095
Internet: www.ppfahome.org

National trade association of manufacturers of plastic piping products used for plumbing applications. Promotes the use of plastic piping products in plumbing applications for water service, water distribution, disposal waste vent, building drainage and sprinkler applications, installed in structures or on premises in accordance with applicable codes.

Plumbing and Drainage Institute
45 Bristol Drive
South Easton, MA 02375
Phone: 800-589-8956
Fax: 508-230-3529
e-mail: info@pdionline.org
Internet: www.pdionline.org

Plumbing-Heating-Cooling Contractors National Association
180 S. Washington St.
P.O. Box 6808
Falls Church, VA 22040
Phone: 703-237-8100
Fax: 703-327-7442
Internet: www.phccweb.org

This organization of approximately 3700 contractors in the USA promotes education and training of the industry.

Southern Building Code Congress International, Inc.
900 Montclair Road
Birmingham, AL 35213-1206
Phone: 205-591-1853
Fax: 205-591-0775
e-mail: webmaster@sbcci.org
Internet: www.sbcci.org

A model code organization which publishes the Standard Building
Code and the Standard Mechanical Code. The SBCC also provides
technical and educational support services.

Water Quality Association
4151 Naperville Rd
Lisle, IL 60532
Phone: 630-505-0160
Fax: 630-505-9637
e-mail: info@mail.wqa.org
Internet: www.wqa.org

The Water Quality Association seeks to assure the right of users of
water to modify or enhance the quality of water to meet specific
needs or desires. Focuses on industry issues, educations and idea
exchange. Commissions various technical studies.

In addition to this abbreviated list, the Engineers Joint Council of
New York publishes a *Directory of Engineering Societies and Related
Organizations* which is available at some local libraries.

GOVERNMENT

United States Environmental Protection Agency
U.S. EPA 4024
401 M. Street NW
Washington, DC 20460
Internet: www.epa.gov

WAVE (Water Alliance To Save Energy)
Water Alliance to Save Energy
Wave Program Director
U.S. EPA Wave Program (4024m)
1200 Pennsylvania Ave., NW
Washington, DC 20460
Phone: 202-564-0623/0624
Fax: 202-501-2396

WAVE is a joint government and industry effort in the hotel and
motel lodging industry to reduce and conserve both energy and
water. WAVE includes many of the largest hotel chains and several
large utilities and manufacturers.

United States Army Corp of Engineers
HQ US Army Corps of Engineers
441 G St., NW
Washington, DC 20314-1000
Phone: 202-761-0000
Internet: www.usace.army.mil

Appendix I
Bibliography of Sources

Avallone, Eugene A. and Baumeister III, Theodore, Editors. *Marks Standard Handbook for Mechanical Engineers*, 9th Edition, New York: McGraw-Hill Publishing Company, 1987.

Bradshaw, Vaughn. *Building Control Systems*, Second Edition, New York: John Wiley & Sons, Inc., 1993.

Building Construction Cost Data, Kingston, MA: R.S. Means Company Inc., 1995.

Burrows, William, Ph.D., *Textbook of Microbiology*, Philadelphia: W.B. Saunders Publishing Company, 1963.

Code of Federal Regulations Chapter 29 (CFR 29 1910), *Occupational Safety and Heath* OSHA Standards.

Code of Federal Regulations Chapter 40 (CFR 40), *EPA Environmental Standards and Laws*.

Cotts, David G., *The Facility Management Handbook*, Saranac Lake, NY: AMACOM, a Division of The American Management Association, 1992.

The Crane Company, *Flow of Fluids through Valves, Fittings and Pipe*, New York, 1979.

Frame J. Davidson, *Managing Projects in Organizations*, San Francisco: Jossey-Wales Co., 1987.

General Accounting Office Report to Congress, *Safe Drinking Water Act: Progress and Future Challenges Implementing the 1996 Amendments*, January 1999. GAO/RCED-00-31

Harris, Cyrill, Ph.D., Editor, *Handbook of Utilities and Services for Buildings: Planning, Design and Installation*, New York: McGraw-Hill Publishing Company, 1990.

Hicks, Tyler G., PE, Editor. *Standard Handbook of Engineering Calculations*, New York: McGraw-Hill Publishing Company, 1972.

Journal, a monthly magazine published by the American Water Works Company, 1995.

Lane, Russell W., *Control of Scale and Corrosion in Building Water Systems*, New York: McGraw-Hill Publishing Company, 1993.

Lyons, Jerry L., PE, and Askland, Jr., *Carl L. Lyons' Encyclopedia of Valves*, New York: Van Nostrand-Reinhold Company, 1975.

Nayyar, Mohninder L., PE, Editor-in-Chief, *Piping Handbook*, 6th Edition, New York: McGraw-Hill Publishing Company, 1992.

Stein, Benjamin and Reynolds, John S., *Mechanical and Electrical Equipment for Buildings*, 8th Edition, New York: John Wiley & Sons, Inc., 1992.

Step by Step Guide Book on Home Plumbing, West Valley City, UT: Step by Step Guide Book Co., 1981.

Tortora, Gerard J., Funke, Berdell R. and Chase, Christine L., *Microbiology: An Introduction*, 4th Edition, Menlo Park, CA: The Benjamin/Cummings Publishing Co. Inc.

Uniform Plumbing Code™, *Safety Requirements for Plumbing*, Walnut, CA: International Association of Plumbing and Mechanical Officials, 1995.

Appendix II

Primary Drinking Water Standards for Community Systems*

Contaminant	MCLG[1] (mg/L)[2]	MCL or TT[1] (mg/L)[2]	Potential health effects from exposure above the MCL	Common sources of contaminant in drinking water
MICROORGANISMS				
Cryptosporidium	zero	TT[3]	Gastrointestinal illness (e.g., diarrhea, vomiting, cramps)	Human and fecal animal waste
Giardia lamblia	zero	TT[3]	Gastrointestinal illness (e.g., diarrhea, vomiting, cramps)	Human and animal fecal waste
Heterotrophic plate count (HPC)	n/a	TT[3]	HPC has no health effects; it is an analytic method used to measure the variety of bacteria that are common in water. The lower the concentration of bacteria in drinking water, the better maintained the water system is.	HPC measures a range of bacteria that are naturally present in the environment
Legionella	zero	TT[3]	Legionnaire's Disease, a type of pneumonia	Found naturally in water; multiplies in heating systems
Total Coliforms (including fecal coliform and *E. coli*)	zero	5.0%[4]	Not a health threat in itself; it is used to indicate whether other potentially harmful bacteria may be present[5]	Coliforms are naturally present in the environment; as well as feces; fecal coliforms and *E. coli* only come from human and animal fecal waste.
Turbidity	n/a	TT[3]	Turbidity is a measure of the cloudiness of water. It is used to indicate water quality and filtration effectiveness (e.g., whether disease-causing organisms are present). Higher turbidity levels are often associated with higher levels of disease-causing microorganisms such as viruses, parasites and some bacteria. These organisms can cause symptoms such as nausea, cramps, diarrhea, and associated headaches.	Soil runoff
Viruses (enteric)	zero	TT[3]	Gastrointestinal illness (e.g., diarrhea, vomiting, cramps)	Human and animal fecal waste
DISINFECTION BYPRODUCTS				
Bromate	zero	0.010	Increased risk of cancer	Byproduct of drinking water disinfection
Chlorite	0.8	1.0	Anemia; infants & young children: nervous system effects	Byproduct of drinking water disinfection
Haloacetic acids (HAA5)	n/a[6]	0.060	Increased risk of cancer	Byproduct of drinking water disinfection
Total Trihalomethanes (TTHMs)	none[7] / n/a[6]	0.10 / 0.080	Liver, kidney or central nervous system problems; increased risk of cancer	Byproduct of drinking water disinfection
DISINFECTANTS	MRDL[1] (mg/L)[2]	MRDL[1] (mg/L)[2]		

*Source: US Environmental Protection Agency

313

Contaminant	MCLG[1] (mg/L)[2]	MCL or TT[1] (mg/L)[2]	Potential health effects from exposure above the MCL	Common sources of contaminant in drinking water
Chloramines (as Cl$_2$)	MRDLG=4	MRDL=4.0[1]	Eye/nose irritation; stomach discomfort, anemia	Water additive used to control microbes
Chlorine (as Cl$_2$)	MRDLG=4	MRDL=4.0[1]	Eye/nose irritation; stomach discomfort	Water additive used to control microbes
Chlorine dioxide (as ClO$_2$)	MRDLG=0 .8[1]	MRDL=0.8[1]	Anemia; infants & young children: nervous system effects	Water additive used to control microbes
INORGANIC CHEMICALS				
Antimony	0.006	0.006	Increase in blood cholesterol; decrease in blood sugar	Discharge from petroleum refineries; fire retardants; ceramics; electronics; solder
Arsenic	0[7]	0.010 as of 1/23/06	Skin damage or problems with circulatory systems, and may have increased risk of getting cancer	Erosion of natural deposits; runoff from orchards, runoff from glass & electronicsproduction wastes
Asbestos (fibers >10 micrometers)	7 million fibers per Liter (MFL)	7 MFL	Increased risk of developing benign intestinal polyps	Decay of asbestos cement in water mains; erosion of natural deposits
Barium	2	2	Increase in blood pressure	Discharge of drilling wastes; discharge from metal refineries; erosion of natural deposits
Beryllium	0.004	0.004	Intestinal lesions	Discharge from metal refineries and coal-burning factories; discharge from electrical, aerospace, and defense industries
Cadmium	0.005	0.005	Kidney damage	Corrosion of galvanized pipes; erosion of natural deposits; discharge from metal refineries; runoff from waste batteries and paints
Chromium (total)	0.1	0.1	Allergic dermatitis	Discharge from steel and pulp mills; erosion of natural deposits
Copper	1.3	TT[8]; Action Level= 1.3	Short term exposure: Gastrointestinal distress Long term exposure: Liver or kidney damage People with Wilson's Disease should consult their personal doctor if the amount of copper in their water exceeds the action level	Corrosion of household plumbing systems; erosion of natural deposits
Cyanide (as free cyanide)	0.2	0.2	Nerve damage or thyroid problems	Discharge from steel/metal factories; discharge from plastic and fertilizer factories
Fluoride	4.0	4.0	Bone disease (pain and tenderness of the bones); Children may get mottled teeth	Water additive which promotes strong teeth; erosion of natural deposits; discharge from fertilizer and aluminum factories
Lead	zero	TT[8]; Action Level= 0.015	Infants and children: Delays in physical or mental development; children could show slight deficits in attention span and learning abilities Adults: Kidney problems; high blood pressure	Corrosion of household plumbing systems; erosion of natural deposits

Contaminant	MCLG[1] (mg/L)[2]	MCL or TT[1] (mg/L)[2]	Potential health effects from exposure above the MCL	Common sources of contaminant in drinking water
Mercury (inorganic)	0.002	0.002	Kidney damage	Erosion of natural deposits; discharge from refineries and factories; runoff from landfills and croplands
Nitrate (measured as Nitrogen)	10	10	Infants below the age of six months who drink water containing nitrate in excess of the MCL could become seriously ill and, if untreated, may die. Symptoms include shortness of breath and blue-baby syndrome.	Runoff from fertilizer use; leaching from septic tanks, sewage; erosion of natural deposits
Nitrite (measured as Nitrogen)	1	1	Infants below the age of six months who drink water containing nitrite in excess of the MCL could become seriously ill and, if untreated, may die. Symptoms include shortness of breath and blue-baby syndrome.	Runoff from fertilizer use; leaching from septic tanks, sewage; erosion of natural deposits
Selenium	0.05	0.05	Hair or fingernail loss; numbness in fingers or toes; circulatory problems	Discharge from petroleum refineries; erosion of natural deposits; discharge from mines
Thallium	0.0005	0.002	Hair loss; changes in blood; kidney, intestine, or liver problems	Leaching from ore-processing sites; discharge from electronics, glass, and drug factories
ORGANIC CHEMICALS				
Acrylamide	zero	TT[9]	Nervous system or blood problems; increased risk of cancer	Added to water during sewage/wastewater treatment
Alachlor	zero	0.002	Eye, liver, kidney or spleen problems; anemia; increased risk of cancer	Runoff from herbicide used on row crops
Atrazine	0.003	0.003	Cardiovascular system or reproductive problems	Runoff from herbicide used on row crops
Benzene	zero	0.005	Anemia; decrease in blood platelets; increased risk of cancer	Discharge from factories; leaching from gas storage tanks and landfills
Benzo(a)pyrene (PAHs)	zero	0.0002	Reproductive difficulties; increased risk of cancer	Leaching from linings of water storage tanks and distribution lines
Carbofuran	0.04	0.04	Problems with blood, nervous system, or reproductive system	Leaching of soil fumigant used on rice and alfalfa
Carbon tetrachloride	zero	0.005	Liver problems; increased risk of cancer	Discharge from chemical plants and other industrial activities
Chlordane	zero	0.002	Liver or nervous system problems; increased risk of cancer	Residue of banned termiticide
Chlorobenzene	0.1	0.1	Liver or kidney problems	Discharge from chemical and agricultural chemical factories
2,4-D	0.07	0.07	Kidney, liver, or adrenal gland problems	Runoff from herbicide used on row crops
Dalapon	0.2	0.2	Minor kidney changes	Runoff from herbicide used on rights of way
1,2-Dibromo-3-chloropropane (DBCP)	zero	0.0002	Reproductive difficulties; increased risk of cancer	Runoff/leaching from soil fumigant used on soybeans, cotton, pineapples, and orchards
o-Dichlorobenzene	0.6	0.6	Liver, kidney, or circulatory system problems	Discharge from industrial chemical factories
p-Dichlorobenzene	0.075	0.075	Anemia; liver, kidney or spleen damage; changes in blood	Discharge from industrial chemical factories

Contaminant	MCLG[1] (mg/L)[2]	MCL or TT[1] (mg/L)[2]	Potential health effects from exposure above the MCL	Common sources of contaminant in drinking water
1,2-Dichloroethane	zero	0.005	Increased risk of cancer	Discharge from industrial chemical factories
1,1-Dichloroethylene	0.007	0.007	Liver problems	Discharge from industrial chemical factories
cis-1,2-Dichloroethylene	0.07	0.07	Liver problems	Discharge from industrial chemical factories
trans-1,2-Dichloroethylene	0.1	0.1	Liver problems	Discharge from industrial chemical factories
Dichloromethane	zero	0.005	Liver problems; increased risk of cancer	Discharge from drug and chemical factories
1,2-Dichloropropane	zero	0.005	Increased risk of cancer	Discharge from industrial chemical factories
Di(2-ethylhexyl) adipate	0.4	0.4	Weight loss, liver problems, or possible reproductive difficulties.	Discharge from chemical factories
Di(2-ethylhexyl) phthalate	zero	0.006	Reproductive difficulties; liver problems; increased risk of cancer	Discharge from rubber and chemical factories
Dinoseb	0.007	0.007	Reproductive difficulties	Runoff from herbicide used on soybeans and vegetables
Dioxin (2,3,7,8-TCDD)	zero	0.00000003	Reproductive difficulties; increased risk of cancer	Emissions from waste incineration and other combustion; discharge from chemical factories
Diquat	0.02	0.02	Cataracts	Runoff from herbicide use
Endothall	0.1	0.1	Stomach and intestinal problems	Runoff from herbicide use
Endrin	0.002	0.002	Liver problems	Residue of banned insecticide
Epichlorohydrin	zero	TT[9]	Increased cancer risk, and over a long period of time, stomach problems	Discharge from industrial chemical factories; an impurity of some water treatment chemicals
Ethylbenzene	0.7	0.7	Liver or kidneys problems	Discharge from petroleum refineries
Ethylene dibromide	zero	0.00005	Problems with liver, stomach, reproductive system, or kidneys; increased risk of cancer	Discharge from petroleum refineries
Glyphosate	0.7	0.7	Kidney problems; reproductive difficulties	Runoff from herbicide use
Heptachlor	zero	0.0004	Liver damage; increased risk of cancer	Residue of banned termiticide
Heptachlor epoxide	zero	0.0002	Liver damage; increased risk of cancer	Breakdown of heptachlor
Hexachlorobenzene	zero	0.001	Liver or kidney problems; reproductive difficulties; increased risk of cancer	Discharge from metal refineries and agricultural chemical factories
Hexachlorocyclopentadiene	0.05	0.05	Kidney or stomach problems	Discharge from chemical factories
Lindane	0.0002	0.0002	Liver or kidney problems	Runoff/leaching from insecticide used on cattle, lumber, gardens
Methoxychlor	0.04	0.04	Reproductive difficulties	Runoff/leaching from insecticide used on fruits, vegetables, alfalfa, livestock
Oxamyl (Vydate)	0.2	0.2	Slight nervous system effects	Runoff/leaching from insecticide used on apples, potatoes, and tomatoes

Contaminant	MCLG[1] (mg/L)[2]	MCL or TT[1] (mg/L)[2]	Potential health effects from exposure above the MCL	Common sources of contaminant in drinking water
Polychlorinated biphenyls (PCBs)	zero	0.0005	Skin changes; thymus gland problems; immune deficiencies; reproductive or nervous system difficulties; increased risk of cancer	Runoff from landfills; discharge of waste chemicals
Pentachlorophenol	zero	0.001	Liver or kidney problems; increased cancer risk	Discharge from wood preserving factories
Picloram	0.5	0.5	Liver problems	Herbicide runoff
Simazine	0.004	0.004	Problems with blood	Herbicide runoff
Styrene	0.1	0.1	Liver, kidney, or circulatory system problems	Discharge from rubber and plastic factories; leaching from landfills
Tetrachloroethylene	zero	0.005	Liver problems; increased risk of cancer	Discharge from factories and dry cleaners
Toluene	1	1	Nervous system, kidney, or liver problems	Discharge from petroleum factories
Toxaphene	zero	0.003	Kidney, liver, or thyroid problems; increased risk of cancer	Runoff/leaching from insecticide used on cotton and cattle
2,4,5-TP (Silvex)	0.05	0.05	Liver problems	Residue of banned herbicide
1,2,4-Trichlorobenzene	0.07	0.07	Changes in adrenal glands	Discharge from textile finishing factories
1,1,1-Trichloroethane	0.20	0.2	Liver, nervous system, or circulatory problems	Discharge from metal degreasing sites and other factories
1,1,2-Trichloroethane	0.003	0.005	Liver, kidney, or immune system problems	Discharge from industrial chemical factories
Trichloroethylene	zero	0.005	Liver problems; increased risk of cancer	Discharge from metal degreasing sites and other factories
Vinyl chloride	zero	0.002	Increased risk of cancer	Leaching from PVC pipes; discharge from plastic factories
Xylenes (total)	10	10	Nervous system damage	Discharge from petroleum factories; discharge from chemical factories
RADIONUCLIDES				
Alpha particles	none[7]	15 picocuries per Liter (pCi/L)	Increased risk of cancer	Erosion of natural deposits of certain minerals that are radioactive and may emit a form of radiation known as alpha radiation
Beta particles and photon emitters	none[7]	4 millirems per year (mrem/yr)	Increased risk of cancer	Decay of natural and man-made deposits of certain minerals that are radioactive and may emit forms of radiation known as photons and beta radiation
Radium 226 and Radium 228 (combined)	none[7]	5 pCi/L	Increased risk of cancer	Erosion of natural deposits
Uranium	zero	30 ug/L as of 12/08/03	Increased risk of cancer, kidney toxicity	Erosion of natural deposits

NOTES

1 - Definitions
- Maximum Contaminant Level Goal (MCLG) - The level of a contaminant in drinking water below which there is no known or expected risk to health. MCLGs allow for a margin of safety and are non-enforceable public health goals.
- Maximum Contaminant Level (MCL) - The highest level of a contaminant that is allowed in drinking water. MCLs are set as close to MCLGs as feasible using the best available treatment technology and taking cost into consideration. MCLs are enforceable standards.
- Maximum Residual Disinfectant Level Goal (MRDLG) - The level of a drinking water disinfectant below which there is no known or expected risk to health. MRDLGs do not reflect the benefits of the use of disinfectants to control microbial contaminants.
- Maximum Residual Disinfectant Level (MRDL) - The highest level of a disinfectant allowed in drinking water. There is convincing evidence that addition of a disinfectant is necessary for control of microbial contaminants.
- Treatment Technique (TT) - A required process intended to reduce the level of a contaminant in drinking water.

2 - Units are in milligrams per liter (mg/L) unless otherwise noted- Milligrams per liter are equivalent to parts per million (ppm).

3 - EPA's surface water treatment rules require systems using surface water or ground water under the direct influence of surface water to (1) disinfect their water, and (2) filter their water or meet criteria for avoiding filtration so that the following contaminants are controlled at the following levels:
- *Cryptosporidium* (as of 1/1/02 for systems serving >10,000 and 1/14/05 for systems serving <10,000) 99% removal.
- *Giardia lamblia:* 99.9% removal/inactivation
- Viruses: 99.99% removal/inactivation
- *Legionella:* No limit, but EPA believes that if *Giardia* and viruses are removed/inactivated, *Legionella will* also be controlled
- Turbidity: At no time can turbidity (cloudiness of water) go above 5 nephelolometric turbidity units (NTU); systems that filter must ensure that the turbidity go no higher than 1 NTU (0.5 NTU for conventional or direct filtration) in at least 95% of the daily samples in any month. As of January 1, 2002, turbidity may never exceed 1 NTU, and must not exceed 0.3 NTU in 95% of daily samples in any month.
- HPC: No more than 500 bacterial colonies per milliliter
- Long Term 1 Enhanced Surface Water Treatment (Effective Date: January 14, 2005); Surface water systems or (GWUDI) systems serving fewer than 10,000 people must comply with the applicable Long Term 1 Enhanced Surface Water Treatment Rule provisions (e.g. turbidity standards, individual filter monitoring, Cryptosporidium removal requirements, updated watershed control requirements for unfiltered systems).
- Filter Backwash Recycling; The Filter Backwash Recycling Rule requires systems that recycle to return specific recycle flows through all processes of the system's existing conventional or direct filtration system or at an alternate location approved by the state.

4 - No more than 5.0% samples total coliform-positive in a month. (For water systems that collect fewer than 40 routine samples per month, no more than one sample can be total coliform-positive per month.) Every sample that has total coliform must be analyzed for either fecal coliforms or *E. coli* if two consecutive TC-positive samples, and one is also positive for *E. coli* fecal coliforms, system has an acute MCL violation.

5 - Fecal coliform and *E. coli* are bacteria whose presence indicates that the water may be contaminated with human or animal wastes. Disease-causing microbes (pathogens) in these wastes can cause diarrhea, cramps, nausea, headaches, or other symptoms. These pathogens may pose a special health risk for infants, young children, and people with severely compromised immune systems.

6 - Although there is no collective MCLG for this contaminant group, there are individual MCLGs for some of the individual contaminants:
- Haloacetic acids: dichloroacetic acid (zero); trichloroacetic acid (0.3 mg/L)
- Trihalomethanes: bromodichloromethane (zero); bromoform (zero); dibromochloromethane (0.06 mg/L)

7 - MCLGs were not established before the 1986 Amendments to the Safe Drinking Water Act. The standard for this contaminant was set prior to 1986. Therefore, there is no MCLG for this contaminant

8 - Lead and copper are regulated by a Treatment Technique that requires systems to control the corrosiveness of their water. If more than 10% of tap water samples exceed the action level, water systems must take additional steps- For copper, the action level is 1.3 mg/L, and for lead is 0.0 15 mg/L.

9 - Each water system must certify, in writing, to the state that when it uses acrylamide and/or epichlorohydrin to treat water, the combination (or product) of dose and monomer level does not exceed the levels specified, as follows: Acrylamide = 0.05% dosed at 1 mg/L (or equivalent); Epichlorohydrin = 0.01% dosed at 20 mg/L (or equivalent).

Index